REFRIGERATION

a practical manual
for apprentices

THIRD EDITION

by

G. H. REED, F.Inst.R.

APPLIED SCIENCE PUBLISHERS LTD
LONDON

APPLIED SCIENCE PUBLISHERS LTD
RIPPLE ROAD, BARKING, ESSEX, ENGLAND

First Edition published 1967
Second Edition, revised, published 1969
Reprinted 1970
Third Edition published 1974
Reprinted 1981

ISBN 0 85334 605 4

WITH TABLES AND 54 ILLUSTRATIONS

© G. H. REED 1967, 1969, 1974

Printed in Great Britain by Galliard (Printers) Ltd Great Yarmouth

Contents

Illustrations

Introduction

The purpose of this book is to provide practical guidance to apprentices entering the refrigeration trade as service engineers. It will of course be of value to anyone wishing to acquire knowledge of refrigeration principles, and function of controls, whose ultimate objective is the installation of equipment and the diagnosing and repairing of faults. The word 'practical' is stressed, because it is not the intention to prescribe large doses of theory, although from time to time, where a theoretical point is essential to the proper understanding of a problem, it will be included. REFRIGERATION with its section of Test Questions provides an excellent textbook for a staff technical training course. Calculations are in SI and Imperial units.

Parts of this book first appeared in the journal *Modern Refrigeration*, now *Refrigeration and Air Conditioning*. Their enthusiastic reception led to many requests for the articles to be made available in permanent form. With this in view, the author has considerably expanded and added to the first articles to make this comprehensive book.

It is not in any way intended to replace theoretical studies, which will always remain an important part of an engineer's training, but rather to be complementary to them. It will deal with aspects of learning to be a refrigeration service engineer, based on experience in the field, which are not normally to be found in any technical syllabus. Sometimes the detail given may appear to be elementary, but so often the basic features are assumed to be known, and therefore neglected in early training, which renders advanced instruction later on more difficult to master.

Refrigeration equipment is continuously being installed for a wide and ever-increasing variety of uses. The obvious ones are the coldrooms to be seen in a butcher's shop, the frozen food cabinets in the supermarkets, and of course the refrigerators for the home, known as the 'domestic refrigerators'. But there are many other applications for refrigeration not so well known, such as air conditioning, cellar cooling, ice-making, blood storage, water chilling, dough retarding and so on, and as all mechanical devices need regular servicing, it follows that, in every locality, the services of a skilled refrigeration engineer are in constant demand. It follows, too, that to be skilled he needs to be conversant with a diversity of machinery. Whilst this contributes to the interest of the work, it also emphasises how much practical knowledge an apprentice has to acquire during his training.

Situated throughout the countryside are a number of distributors or dealers, whose aim it is to supply the demand for refrigeration machinery, and to maintain it in efficient working order afterwards. These companies vary in size, but an average one might consist of a service manager, six engineers (complete with vans), two apprentices, a storekeeper, and perhaps a delivery man. It will have a workshop, a showroom, and of course sales and office staffs, and this is the type of company I have mainly in mind in this book.

Apprentices in other spheres of refrigeration, manufacturing for example, will no doubt be expected to concentrate more on the technique of design and construction, but the service engineer apprentice is concerned with the installation of plant, with diagnosing and repairing faults. It is men entering this profession that these instructions are intended to assist.

Before proceeding to practical matters, I would like to offer a few words of guidance, which I hope will be of assistance to the apprentice beginning the 'first day' with his company.

Not only will it be the first day of a first job, but transition from school to employment; an important stage in everyone's career. Comfort may be found in knowing that for most people 'first days' never lose their tension.

So to the new recruit I would say this:

Be individualistic and think for yourself. By all means be ready to listen to advice and instruction, and give due consideration to it, but having listened, think and form your own opinion.

Be decisive. Refrigeration service demands resourcefulness, for so often the engineer has to make his own decisions and act accordingly.

Be tidy, respectful, trustworthy, and acquire knowledge, not so much to satisfy your employers, but for the sake of your own pride.

Be on time in the mornings. Punctuality indicates a respect for time, and a sense of responsibility; and the good impression it creates is a reward far in excess of the effort it requires.

Finally, learn for yourself. It is natural after ten or eleven years at school, with a teacher giving instructions, saying how and when a task should be done, and checking it afterwards, to have developed an attitude of 'expecting to be taught'. Those days are over. Now the questions must be sought, the answers found. No matter how helpful an engineer may wish to be, it must be remembered that he has a job to do. In other words, maturity brings its own obligations, and finding out what to do, rather than waiting to be told, is one of them.

If these points can be appreciated from the 'first day' it will indeed have been a valuable one.

1
Tools and a Tool-box

Tools, and a box to put them in, are essential equipment to any tradesman, and a refrigeration apprentice should certainly acquire a tool-box at the earliest opportunity. Not only does its possession establish a sense of status, but it provides a proper place for the tools he will require, and which can be purchased gradually.

Tool-boxes of all shapes and sizes can be bought ready made, both in wood or metal, and my own preference is for one in wood. For one thing wood does not rust, it absorbs oil and damp, and if strongly made, withstands rather better the general wear and tear inevitable in the life of a refrigeration engineer. A useful box can quite easily be made by those who like to do things for themselves, and save some money in the process.

Figure 1 is offered as a guide, the important features of which are the length to accommodate gauge lines, that it is fairly tall and narrow to facilitate carrying, and it has a tray fitted into the top to hold the tools most frequently used.

The drawing is self-explanatory, and the dimensions obviously may be modified if required. At least 10 mm plywood is suggested, and this should be accurately cut and prepared, using a set square to ensure right-angles. Dovetailing of the sides would be advocated, but if this is beyond the capabilities of the constructor, glue and screws will make an effective substitute. The lid should be fitted with two strong brass hinges, and a hasp and staple piece, which can then be secured with a padlock. Both hinges and hasp should be fixed with small nuts and bolts, rather than screws. The tray protrudes above the box level by about 12 mm, and this serves to support the lid when shut. Two pieces of cord, threaded through two holes either side, with a knot on the four ends inside, will, as shown, provide a means of carrying the box. A coat or two of wood dye will complete the job. Figure 2 shows the completed tool-box.

More important than the box, of course, are the tools, which are part of the tradesman himself. They are the mechanical link between

1

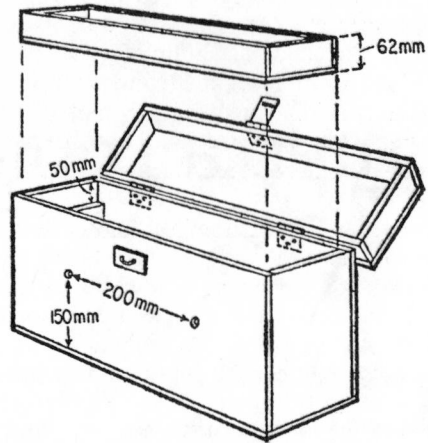

Fig. 1.   A useful tool-box which an apprentice can make quite easily.

his knowledge and the machine; an extension of his hands as it were, without which he can be powerless. It follows therefore that the more effective the link, the more efficient and easier will be the work. It follows, too, that tools being the tradesman's servants, it is his duty as master to look after them. They are not cheap, but fortunately tools are items which can be obtained piece by piece, and the following is a list placed in the order in which it is advised they be purchased.

PAIR OF PLIERS, AND SMALL ELECTRICAL SCREWDRIVER, WITH NEON TESTER

An engineer will frequently require his assistant to fit a lead to a plug top, or similar work involving cable, and these two simple

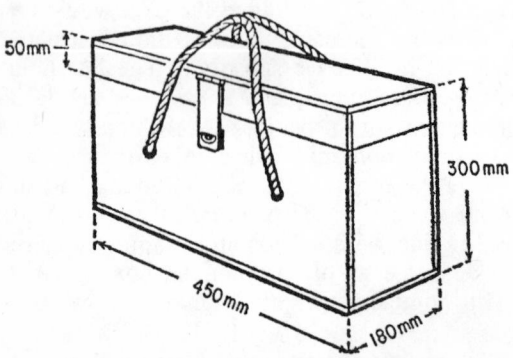

Fig. 2.   The completed tool-box.

tools will be of immense value. The additions of a neon tester in the screwdriver handle, and insulation on the pliers, are added advantages. The first to discover if a connection is 'live', and the second as protection if this precaution is not observed. Electric shocks are anything but pleasant, and can be dangerous—even fatal; nevertheless, despite repeated warnings, the best tutor seems to be experience.

Pliers need to be sharp to cut the lead and covering cleanly, and if a fair amount of electrical work is undertaken, the addition of side cutters and insulation strippers is advised, the latter for paring the insulation on the flex. Care must be taken not to cut any of the strands during this operation. Not only is the lead weakened if this is done, but its current carrying capacity is reduced.

SELECTION OF SCREWDRIVERS

Screwdrivers of all sizes are invaluable. Always have the correct size driver for the screw to be removed or inserted. A too small one will slip, and once this happens, damage to the screw head may be such as to render the proper size ineffective. Short stubby handles are indispensable in a confined space, just as long handles are essential for long range operation. Remember, too, that screwdrivers are not chisels, and should never be hit with a hammer, or otherwise abused.

'PHILLIPS' SCREWDRIVERS

'Phillips' screws are widely used nowadays; they are the heads with a 'cross'. The 'Roberts', found mainly on Canadian cabinets, are those with a 'square', and both are practically immovable without the corresponding driver of the correct size. These heads are illustrated in Fig. 3.

Fig. 3. Standard, Phillips and Roberts screw heads. It is essential to use the correct screwdrivers for these screws.

ALLEN KEYS

For undoing grub screws on fan blades and motor pulleys.

ADJUSTABLE SPANNER

Adjustable spanners (*see* Fig. 4), of which there are several versions, are very useful instruments, and will prove themselves good friends on many occasions, but for all that, use them only as second best to a proper sized conventional spanner. They have a tendency to slacken and slip in use, causing damage either to fingers or nut—or both! There is a right and a wrong direction to use them and Fig. 4 illustrates this.

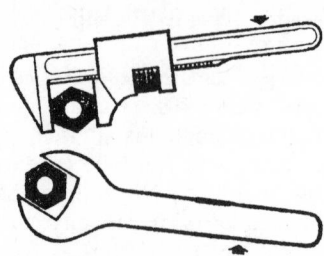

Fig. 4.   Be sure to use your spanners in the correct direction as shown.

OIL CAN (WITH SCREWED CAP AND PUMP ACTION)

A pump action is necessary otherwise it will be quite impossible to use where space above the motor bearings to be oiled is limited. Also a flexible nozzle is a valuable refinement, and of great assistance at such times. A nozzle cap, apart from preventing dirt entering the nozzle, stops oil leaking into the tool-box when not in use.

HACKSAW, JUNIOR HACKSAW

The type of hacksaw recommended, adjustable for varying blade lengths, is illustrated in Fig. 5, and the correct direction of the blade is shown. A general purpose blade has 18 teeth to the inch. The 'wavy' edge blade is preferable for soft metals, and it is advisable to carry a selection of spare blades. A junior hacksaw is a useful accessory to deal with the inevitable space difficulties.

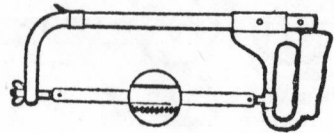

Fig. 5.   Hacksaw—this is the type recommended. Note the correct direction of the blade as shown in the circled section.

HAMMER (0·5 kg), HAMMER (0·125 kg)

A hammer has innumerable obvious uses, but it is a tool to be used sparingly and with discretion. Much extra work is often caused by heavy-handed efforts with this instrument.

UNIVERSAL PLIERS
A useful tool for round objects.

SET WHITWORTH SPANNERS, SET SAE SPANNERS
Open-jawed spanners, despite their many modern rivals, still play an important role in any tool kit, since they can often perform their function where more sophisticated spanners refuse to operate. British Whitworth Standard, and American and Unified Standard (SAE), are the two most common types to be found in refrigeration service work. These spanners should also be used in the direction as shown in Fig. 4.

STEEL RULE
A steel rule naturally has many uses, and the only advice that seems necessary is to stress the need for accuracy. It is true that a service engineer will seldom be called upon to work to the fine precision of a machine shop engineer, nevertheless pride in absolute accuracy must be developed.

ELECTRICAL TEST LEAD
Described on p. 74.

SCISSORS, FLAT FILE, KNIFE
The knife and a pair of scissors may not strictly speaking be classified as tools, but they are very useful allies to possess. The file is required to assist in the making of 'flares' on copper tubing, a task apprentices will frequently be requested to do.

A more advanced list of tools is suggested as follows: bullnose pliers, tin snips, wire cutters, ring spanners, set of box spanners with

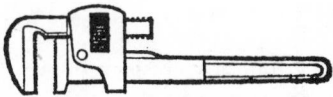

Fig. 6. Stilson. This is an indispensable tool for screwed pipework.

ratchet, medium-size stilsons (Fig. 6), selection of 'Roberts' screwdrivers, selection of files, including a small 'rat-tail' and a large round. Hand drill complete with drills, centre pop, punch, cold chisels, inside and outside calipers, hollow punches, electric soldering iron.

This second list is really an extension of the first. Regarding ring spanners, these are highly efficient and the leverage possible is often greater than the nut can withstand. Care when using them is therefore advised, since much unnecessary labour can be caused by shearing a stud.

Files of different sizes and shapes are always of service, and a small 'rat-tail' or 'needle round' is of particular value in elongating small holes in metal such as are often required to adapt a thermostat plate.

A Stilson, illustrated in Fig. 6, comes to some extent within the classification of an adjustable spanner, but its special and indispensable purpose is dealing with pipe work on water-cooled plants.

There is really no end to a list of useful tools, and experience will soon reveal any deficiencies. For example, a short open-jawed spanner is generally of far more use than a long one, or a ring spanner, when removing the bolts from an open unit motor base, where movement is frequently limited. The point is that the correct size of spanner is not enough, it must be the right *kind* of spanner. A good engineer will see to it that he has the right kind, the poor one will go on wasting time and effort with inadequate tools.

In addition to the above, a skilled refrigeration man will require the following special tools: flare block, pressure gauge, compound pressure gauge, flexible gauge lines, brass elbows (adaptors), flare spanners ($\frac{1}{4}$ to $\frac{3}{4}$), ratchet spanner (with extensions), gland keys, crimping tool, test lamp, thermometer, tube cutter.

EXTRAS

A service van will be supplied with spare parts, which will include refrigerant, oil, paraffin, methylated spirits, carbon tetrachloride, cleaning rag, and a few additional items to include are as follows: selection of the more common type of head plate reeds and gaskets (stored in an air-tight tin). A piece of plate glass about a foot square. A few sheets of medium and fine 'wet and dry' carborundum paper, the fine being about number 500. Metal polish, fuse wire, insulating tape, grease, a few small price labels (*see* p. 89), light lead.

Armed now with a tool-box, an idea of the type of tools to put into it, plus a little knowledge of their uses, one final word. Learn to look after them, keep them clean, watch the chap who is always on the borrow, and above all, get into the habit of counting them before leaving a job.

2
Dirt and Moisture

There are two serious enemies of refrigeration, against which every serviceman must maintain constant vigilance. They are *dirt* and *moisture*.

It is because so many breakdowns are caused by these two items being allowed to enter a refrigeration system, and because it is of the utmost importance that every apprentice should be aware of this danger, that they are mentioned at this early stage.

The amount of dirt or moisture may be very small, and therein lies the difficulty of detection. It can enter a system unnoticed, and the first indication of its presence is a mechanical failure. Cleanliness and elimination of air are absolutely vital in refrigeration service, and this cannot be over-emphasised.

All forms of engineering demand a high degree of cleanliness to maintain efficiency, but a wash through with petrol or some similar fluid is often considered to be adequate. Assembled now with fresh oil, the machine will operate quite satisfactorily, even if the cleansing has not been too thorough. Moreover, we have all seen sparking plugs or other components removed from a combustion engine, when air and possibly specks of dirt have entered the cylinders, without any obvious ill effects.

This is just not good enough for a refrigerator compressor. It demands and must have a much higher standard. In the factory where the compressor or system is originally assembled, stringent attention to cleanliness is observed, using scouring plant, de-greasing fluids, clean hands, with assembly in an air-conditioned room. Finally, as even conditioned air contains moisture, the whole is dried out by heating under a vacuum, the vacuum being broken by allowing dry nitrogen or refrigerant into the system. Thus every precaution is taken in the manufacturing stage to eliminate dirt and moisture, and it is the duty of the service engineer to see that this standard is preserved.

DIRT

For the benefit of the man still at the beginning of his training, it may be of value to explain that specks of dirt or dust, if allowed into the system, can quite easily be deposited on to the seats of compressor valves, thus preventing the rather delicate reeds from seating properly. The result of this is to reduce efficiency, or obstruct pumping altogether.

MOISTURE

Moisture, similarly, causes difficulties out of all proportion to the quantity allowed into the system. When talking of moisture, we are thinking of microscopic quantities, such as the dampness that blowing through a pipe might deposit. The most common danger is air. Water is suspended in the atmosphere, and quite apart from any problems introduced through mixing air with refrigerant, the moisture will almost certainly gather at the expansion valve, where it will form into a pellet of ice, blocking the orifice and so stopping refrigeration.

Since the manufacturer and all good installation engineers take so much trouble to eliminate dirt and moisture, it seems clear that most of the difficulties arising from these factors do so through the opening up of the system after it has been put into use. This may be simply through a normal mechanical failure, *e.g.* a fractured pipe or burst shaft seal allowing refrigerant to escape. This is an accepted hazard, and one of the reasons a service engineer is necessary. What has to be guarded against is the possibility of dirt or air being allowed into a system, merely because gauges have been fitted for some comparatively minor reason.

Having dealt briefly with the results of dirt and moisture, and stressed the importance of prevention, the apprentice must quickly acquire knowledge of the elementary precautions that every refrigeration service engineer must take in his work.

Gauge and charging lines are obvious dangers. One shudders at the thought of many lines, at this very moment, lying in some grubby tool-box, unplugged and open to the atmosphere, jostled amongst sundry tools, to be produced shortly to charge a plant. The worst feature, perhaps, is that the results of this negligence may not become apparent for some weeks, by which time the culprit will not be suspected. When the expansion valve blocks, or the compressor ceases to function, the unfortunate owner will curse his machinery and send for the engineer again. Let us hope, for his sake, that he does not get the same one.

PREVENTION

The first lesson in prevention, therefore, is always to plug all lines tightly immediately after use (*see* Fig. 7), and on removing the plugs put them into a clean tin kept for the purpose, and always purge a little refrigerant through the lines before opening up to the system.

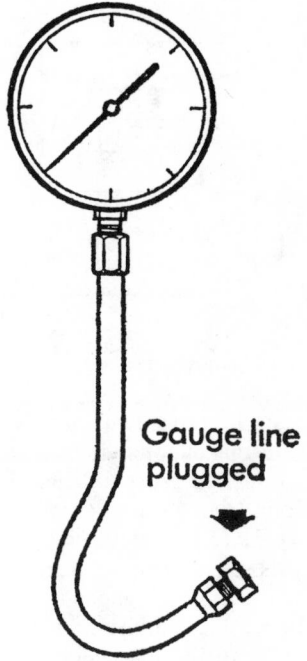

Gauge line plugged

Fig. 7. This is how you should plug your gauge lines immediately after use.

Should it become necessary to expose part of a plant to the atmosphere at any time, it must be blanked off, plugged, or otherwise protected from ingress of air, unless the operation can be carried out quickly. This is particularly important when changing an expansion valve, which may well be wet as the result of defrosting.

REMEDY

Everything should be got ready before removing the valve, including extensive drying operations on and around it, so that no time is wasted in refitting.

Gas cylinder connections, too, must always be blanked off after use (Fig. 8). Not only does this preserve the clean standards demanded of refrigeration, but it also prevents gas escaping should the cylinder shut off valve leak.

When a length of copper tubing is cut from a new coil, the end of the coil should not be left open to the atmosphere, but immediately hammered flat to seal it.

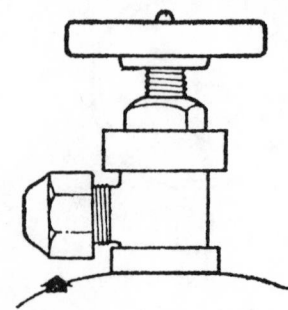

Fig. 8.   Gas cylinder valve. The valve must be capped as shown when not in use.

Perhaps to the very new recruit, items with which he is unfamiliar have been mentioned in this chapter. It is intended to discuss these in greater detail later on, but I trust sufficient has been said, at this stage, to impress upon the apprentice the prime necessity for scrupulous cleanliness. It is his responsibility to carry on 'in the field' the high standards set by the manufacturer.

Breakdowns can so easily be engendered by carelessness or ignorance, resulting in repeated failures, which are a costly nuisance to the owner, frustration to the manufacturer, and embarrassment to the service department. Remember it is the duty of an engineer to prevent problems, just as much as to overcome them.

3
Assisting the Engineer

The objective of an enterprising apprentice should be to make himself useful as soon as possible. He can be of great assistance to an engineer, but this usefulness will be in proportion to his own ability and readiness to learn. As mentioned before, he must seek his knowledge, learn from observation, ask questions, and in this way will progress rapidly. Moreover he must realise there is more to refrigeration service than technical know-how, important though that is.

In the first place, refrigeration is providing an essential service to the public, whether it be concerned with air conditioning in an expensive restaurant, a frozen food cabinet in the modern store, or an ice cream conservator in the little shop round the corner. Refrigeration is not a luxury, but an economic necessity in business, and any breakdown must receive prompt and efficient attention. The engineers are the men in constant contact with the customer; in fact, in many instances they will be the only people the customer knows. The refrigeration company, therefore will be judged, quite unconsciously perhaps, by the manners, appearance and general attitude of the engineer, and this applies equally to the apprentice who accompanies him.

He must take note and appreciate this aspect, therefore, and always be tidy, wear clean overalls, and as far as possible have clean hands. He must be polite, not smoke in front of customers unless invited to do so by them, and if given a cup of tea, the empty cup should be returned, washed if possible, with a word of thanks. Gestures like this do much to create and maintain the goodwill of customers. Above all, do not stand about doing nothing. There is always something to be done, even if it is merely a pretence of activity.

With these generalities as a background, and to illustrate in greater detail the way in which an apprentice can assist, let us follow him on an imaginary journey setting off with an engineer in the morning. They have a quota of, say, five or six service calls, and whilst the driver is manoeuvring through the streets, the apprentice can find a useful occupation by arranging service sheets into the order suggested by the engineer, or discussing the history of the calls. He can, by consulting a street map, give directions on how to reach the first address, or ask the way if necessary. He can also instruct on parking, or give similar guidance, for in these days of traffic congestion, any help a driver can obtain in this way saves time. The apprentice must not think that when travelling from one place to another he need only relax and admire the scenery.

On arriving at their destination he can carry the engineer's tool-box, and should soon be familiar enough with the tools to anticipate the one likely to be required by watching the progress of the work, and in so doing understand what is going on. Should refrigerant be necessary, he will be expected to collect the cylinder from the van, and this applies to all items not normally carried in the tool-box, as, for example, a leak detector.

TEST LAMPS

Lighting a test lamp is something every apprentice must learn to do, and although the engineer will no doubt give instructions, the following points will be of value. I am referring to the old type 'Tilley' leak detector, which in principle is similar to a blow lamp, and which is still popular despite its many rivals. Sometimes it is somewhat temperamental in starting, but the thing to remember here is to be patient, and give the apparatus plenty of time for pre-heating. Always use a wind-shield, and even then select a sheltered spot in which to light it. Do not 'pump up' too soon, otherwise the flame is liable to flare up, and for this reason it is advisable to prepare the lamp in a safe place, away from anything inflammable.

When the vaporiser is fully heated—and it may take two soakings of methylated spirits to do this—give one or two pumps only to start with, waiting until the flame is even and steady before increasing the pressure by six or seven actions. Excessive pressure is not necessary.

Once lit, these lamps are both efficient and safe, if handled with ordinary common sense. It can be seen, however, that they do take some while to prepare, which is valuable time saved as far as the engineer is concerned, if the apprentice can manage this for him.

FLARING PIPES

Making a 'flare' on copper pipe is also within the capabilities of an apprentice, and here again the engineer will provide a demonstration.

The main point to observe is the preparation of the tube end. This must be cleanly cut, preferably by means of a pipe-cutter with a sharp wheel, reamered on the inside to remove any rough edges, and the end smoothed off with a small fine file. In these operations every care must be taken to prevent filings, or other foreign matter, from entering the tube, and where possible the tube should be prepared with the opening downward. Figure 9 shows a flare-block with copper tube firmly clamped ready for flaring. The end

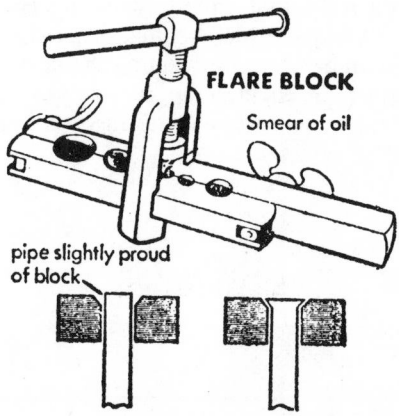

Fig. 9. Making a 'flare'. Sectional view of flare-block showing prepared copper tube end in position before and after using the flaring tool.

must not protrude too far, about 3 mm, and after lightly oiling the flare tool at the pressure point (and ensuring the flare nut is on), it can be tightened up. On old copper, which may have become hard and brittle, it is necessary to anneal the end by heating to cherry red and then quickly quenching it in water, thus preventing the pipe end from splitting during the flaring operation. It is important to apply gentle heat afterwards to evaporate any moisture remaining in the tube end.

Condensing units are frequently sited remote from the cabinet they are refrigerating. In hotels or supermarket installations, for example, the unit may be located on the roof of the building, and even in the average butcher's shop it may be awkwardly situated on top of the coldroom. In such cases the apprentice is of immense value when it comes to checking temperatures, switching the plant on and off as requested, or passing up the tools. The point is he must remember to be within call of the engineer when needed.

With the repair completed, he can assist the engineer in making up the service sheet, drawing his attention to unit numbers (so often forgotten), for it must be realised that a concise account of the work done is an important part of the job itself. From the information the engineer provides an invoice will eventually be presented, and it is also a permanent record to be filed for future reference. Making the sheet up on site, while the facts are still fresh, is an important point to stress, and finally obtaining the customer's signature.

Finally, the apprentice can help by collecting the tools and checking them, putting any old parts aside for exchanging at the store, and in having a general tidy up. Before leaving he should acquire the habit of taking a last look round, the thermometer in particular being the item most frequently left behind, then requesting permission to wash hands in preparation for meeting the next client.

In this way a service engineer apprentice will quickly learn the practical side of his trade, and be of real assistance to those already engaged in it.

4

First Principles

The student might quite reasonably think of refrigeration as being 'cold', but like 'warm' and 'hot', 'cold' is a relative term only. Bath water is often said to be 'cold', whereas if the same temperature were experienced in the sea it would be considered 'warm'.

HEAT

'Cold' cannot be measured, and it is essential in any form of engineering to deal with measurable facts. In refrigeration, strange as it may seem, we are dealing with 'heat', and to make something 'cold' we must extract heat. The degree of heat can be ascertained by means of a thermometer. This is an instrument, no doubt familiar to everyone, consisting of a glass stem, with a fine bore, and a reservoir or bulb at the bottom. In this is contained mercury or alcohol, both of which are liquids with a high coefficient of expansion, the space above the liquid having been exhausted of air before the top of the stem is sealed. Exposed to temperature changes the liquid will expand or contract, rising or falling in the stem as a result.

MEASUREMENTS

At present there are two systems of measurement in general use—the Imperial system used in Britain and America, which includes pounds, feet, inches, etc., and the metric system used in most other countries, which is based on the kilogram (about 2¼ lb) and the metre (just over a yard).

SI UNITS (SYSTÈME INTERNATIONAL)

A completely new system has now been introduced, the object of which is to rationalise measurements and thereby make calculations simpler. These new measurements are based largely on the metric system, although differing in certain respects.

It is likely that for some years both Imperial and SI units will be used in Britain, and for this reason calculations in both systems are given in this book. Naturally, these are confined to the relatively simple examples given, and any student not familiar with SI units should make a special study of the subject. Most people find SI easier than Imperial when they are used to it (*see* p. 145 for notes on SI units).

TEMPERATURE

The Imperial system uses the Fahrenheit scale. The SI system uses the Celsius (formerly Centigrade) scale. Both are related to the freezing and boiling points of water. On the Fahrenheit scale water freezes at 32° and boils at 212°. On the Celsius scale it freezes at 0° and boils at 100° (*see* Fig. 10).

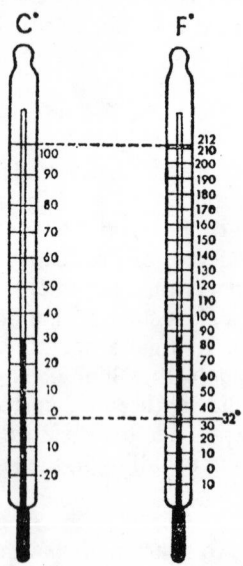

Fig. 10. Centigrade (Celsius) and Fahrenheit thermometers.

It is perhaps necessary to emphasise that these temperatures only apply at normal atmospheric pressures, and that the boiling and freezing points of any liquid may be varied by increasing or decreasing the pressure exerted upon it.

If we look at a thermometer marked with both scales, we can see that 5 Celsius degrees occupy the same length of stem as 9 Fahrenheit degrees. Furthermore, if our thermometer is a low temperature one, we can see that −40°F and −40°C are at the same level on the stem. These two facts give us an easy way to convert Fahrenheit to Celsius and vice versa.

To convert Fahrenheit to Celsius:
1. Add 40
2. Multiply by $\frac{5}{9}$
3. Subtract 40

Example:

Convert 68°F to °C
1. $68 + 40 = 108$
2. $108 \times \frac{5}{9} = 60$
3. $60 - 40 = 20$

68°F = 20°C

To convert Celsius to Fahrenheit:
1. Add 40
2. Multiply by $\frac{9}{5}$
3. Subtract 40

Example:

Convert 35°C to °F
1. $35 + 40 = 75$
2. $75 \times \frac{9}{5} = 135$
3. $135 - 40 = 95$

35°C = 95°F

Combining the two rules we have:

To convert Celsius degrees into Fahrenheit degrees or vice versa:

1. Add 40.
2. Multiply by $\frac{9}{5}$ or $\frac{5}{9}$
3. Subtract 40.

It is easy to decide whether to multiply by 9/5 or 5/9 if it is remembered that °F are packed closer together so there must be more of them.

QUANTITY OF HEAT

A thermometer will record the *degree* of heat, but it will not tell us the *quantity* of heat contained in a substance. Let us consider a glass of hot water and a bucket of hot water at the same temperature. The *quantity* of heat contained in the bucket will be much greater than that contained in the glass. Proof of this may be demonstrated by mixing a glass of hot water and a glass of cold water together—the final temperature will be about midway between the two starting temperatures. Now add a glass of cold water to a bucket of hot water—the final temperature will be much higher than the first mixture.

HEAT CONTENT

Btu AND kJ

Heat content is measured in British thermal units (Btu) in the Imperial system and in kilojoules (kJ) in the SI system. Whilst we have seen that it is possible to measure temperature direct by means of a thermometer, the Btu or kJ can only be calculated from known facts. We can, however, see the result of heat being given to or taken from a substance by observing its change in temperature.

A Btu may be defined as:

The quantity of heat required to be given to or taken from 1 *lb of water in raising or lowering its temperature by* 1°F.

A kJ, which is used for measuring all forms of energy, can be defined in various ways, such as:

A force of 1 *kilonewton acting through a distance of* 1 *metre*

or:

An expenditure of 1 *kilowatt of power for* 1 *second.*

A kJ is slightly smaller than a Btu, and for comparison it takes 4·18 kJ to increase the temperature of 1 kg of water by 1°C.

KILOCALORIE

The student may well meet a third unit, the kilocalorie (kcal), in connection with heat content. This is part of the old metric system and may be defined as:

The quantity of heat to be given to or taken from 1 *kilogram of water in raising or lowering its temperature by* 1°C.

It follows therefore that:

$$1 \text{ kcal} = 4·18 \text{ kJ}$$

1 kcal is equal to 3·97 Btu, which provides the following useful comparison:

$$1 \text{ kcal} = 3·97 \text{ Btu} = 4·18 \text{ kJ}$$
$$1 \text{ Btu} = 0·252 \text{ kcal} = 1·053 \text{ kJ}$$
$$1 \text{ kJ} = 0·24 \text{ kcal} = 0·95 \text{ Btu}$$

The student should, whenever possible, avoid conversion from one system to another by working in the units in which the problem is given.

From the heat units given, consider the following examples.

Imperial Units
Example 1

How many Btu are required to raise the temperature of 1 lb of water from 60°F to 100°F?

$$\text{Btu} = m \times s \times t \qquad \text{where } m = \text{Mass of substance}$$
$$s = \text{Specific heat}$$
$$t = \text{Temperature rise}$$

$$\text{Mass of water} = 1$$
$$\text{Specific heat of water} = 1$$
$$\text{Temperature rise} = 100 - 60 = 40$$
$$\text{Btu} = 1 \times 1 \times 40$$
$$= 40 \text{ Btu}$$

Example 2

How many Btu are required to raise the temperature of 12 lb of water from 50°F to boiling point?

$$\text{Btu} = m \times s \times t \qquad m = 12$$
$$s = 1$$
$$t = 212 - 50 = 162$$
$$= 12 \times 1 \times 162$$
$$= 1\,944 \text{ Btu}$$

Example 3

It is required to reduce the temperature of 24 lb of water from 120°F to 40°F. How many Btu must be extracted?

$$\text{Btu} = m \times s \times t \qquad m = 24$$
$$s = 1$$
$$t = 120 - 40 = 80$$
$$= 24 \times 1 \times 80$$
$$= 1\,920 \text{ Btu}$$

SI Units
Example 1

How many kJ are required to raise the temperature of 1 kg of water from 20°C to 50°C?

$$\text{kJ} = m \times s \times t \qquad m = 1$$
$$s = 4{\cdot}18$$
$$t = 50 - 20 = 30$$

To raise 1 kg by 1°C requires 4·18 kJ.

$$kJ = 1 \times 4·18 \times 30$$
$$= 125·4 \text{ kJ}$$

Example 2

How many kJ are required to raise the temperature of 5 kg of water from 10°C to boiling point?

$$kJ = m \times s \times t \quad m = 5$$
$$s = 4·18$$
$$t = 100 - 10 = 90$$

$$kJ = 5 \times 4·18 \times 90$$
$$= 1\ 881 \text{ kJ}$$

Example 3

How many kJ must be extracted to reduce the temperature of 10 kg of water from 50°C to 10°C?

$$kJ = m \times s \times t \quad m = 10$$
$$s = 4·18$$
$$t = 50 - 10 = 40$$

$$kJ = 10 \times 4·18 \times 40$$
$$= 1\ 672 \text{ kJ}$$

SENSIBLE HEAT/LATENT HEAT

When heat can be seen to be given to or taken from a substance as recorded by a thermometer, it is known as 'sensible heat' (heat which can be 'sensed'). But consider for a moment a pan containing a quantity of ice, the temperature of which is known to be 0°C (32°F). If heat is applied to the pan the ice will melt into water but the temperature will remain at 0°C (32°F). It would be reasonable to ask where the heat applied to the ice has gone. The answer is that it has been used to change the *state* of the ice from a solid to a liquid, and this is known as the 'latent heat of fusion' (*i.e.* 'hidden' heat).

Further, consider the water still at 0°C (32°F) and heat still being applied. The temperature will rise and can be recorded, providing evidence of the sensible heat already mentioned. At 100°C (212°F) (boiling point) the water begins to change its state again, this time from a liquid to a vapour, and during this period the temperature again remains constant. The heat being used for this purpose is known as the 'latent heat of evaporation'.

LATENT HEAT

Latent heat may be defined as:

The heat given to or taken from a substance to change its state without changing its temperature.

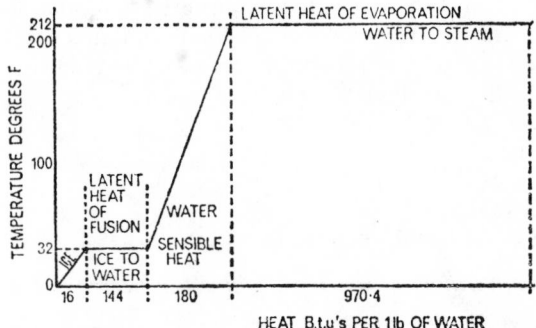

Fig. 11. This graph illustrates the number of Btu required to transfer 1 lb of water at 0°F into 1 lb of steam at 212°F. Note the latent heat stages where no change of temperature takes place.

Figures 11 and 12 illustrate by means of simple graphs the process just described, together with the actual number of Btu or kJ required in each stage. It is important to understand this fully, and to appreciate that *cooling* the 1 lb of steam to 1 lb of ice, or the 1 kg of steam to 1 kg of ice, would require an equal amount of Btu or kJ to be extracted.

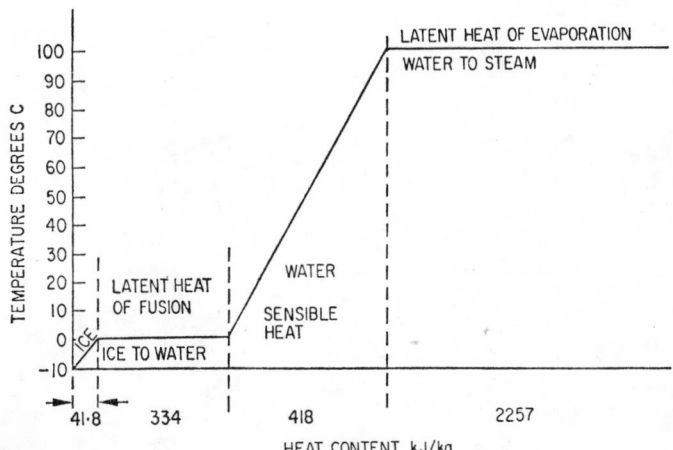

Fig. 12. Graph showing kilojoules required to transfer 1 kg of water at −20°C into 1 kg of steam at 100°C.

It is to be remembered too that all substances can be represented as solids, liquids or gases, by varying the surrounding temperature or pressure. Water is the most familiar substance with which to illustrate this point: ice cubes in a refrigerator, where the temperature has been reduced, and steam escaping from a kettle after heating to boiling point. Where such conditions exist naturally, for example at the North Pole, fresh water is not found in liquid form, it will be ice, and if water were introduced into a room permanently heated to 100°C (212°F) it would evaporate into steam.

Water only remains in liquid form by virtue of the conditions surrounding it.

BOILING POINT
Boiling Point may be defined as:

The temperature at which a liquid is transferred into a vapour.

It must be stressed that this temperature is not necessarily high, as in the case of water. Many liquids, as for example refrigerants, boil well below the freezing point of water.

Just as conditions can be created to change the state of water, so conditions can be created to change the state of refrigerants.

The basis of mechanical refrigeration is the latent heat required to turn a liquid refrigerant into a vapour at its low boiling point.

5
The Refrigeration Cycle

ATMOSPHERIC PRESSURE

In the last chapter, when referring to the freezing and boiling points of water, it was stressed that the temperature stated applied at atmospheric pressure only.

Most people probably clearly understand atmospheric pressure (AP), but for the benefit of those who do not, and to maintain a logical sequence of facts, it must be explained that the earth is blanketed by the air which we breathe. At sea level the pressure exerted by the weight of this air is approximately 14·7 lb/in^2 or, as measured in SI units, approximately 1 bar, but at a higher level it is less dense, and therefore the pressure is lower. In these days of high-flying aircraft, AP has become a familiar feature in our lives, where the plane interior has to be pressurised to preserve comfortable conditions for both crew and passengers, and spacemen, rocketed out beyond the blanket of air, in their capsules, have to wear special suits.

AP has very little to do with refrigeration, but it must be understood in order to appreciate what is meant by a vacuum (which will be dealt with later) and to illustrate how variations of pressure affect the boiling point of a liquid.

Consider now a container of water, to which heat has been applied until the water is evaporating into steam at 100°C (212°F). If the vapour is prevented from escaping, the steam pressure will build up, and as the pressure increases so does the temperature of the water, until, by the time the pressure has doubled, the boiling temperature will have risen to 121°C (250°F). If the heat being applied is removed, and the temperature allowed to drop, the steam will gradually condense back into water, even though the temperature is well above the boiling point of water at atmospheric pressure. This fact can be demonstrated by allowing water to escape from a boiler whilst under pressure. As the water enters the reduced pressure of the atmosphere, it boils off into steam, giving the appearance of steam escaping.

This experiment proves two points:

1. *That boiling point of a liquid can be raised by exerting a pressure upon it.*
2. *That a vapour or gas may be liquefied at a temperature above its atmospheric boiling point, if it is cooled whilst under a pressure.*

With the principles so far discussed it should now be possible to understand the process of refrigeration, and to assist in this, consider a common refrigerant like R.12, contained in a cylinder as carried by the engineers on their vans. Although referred to as 'gas cylinders', for the most part the refrigerant is in the form of a liquid.

Now at atmospheric pressure, R.12 boils at $-29.5°C$ ($-21°F$) ($53°F$ below the freezing point of water at AP), but in the cylinder the liquid refrigerant has only partly boiled (taking heat from the cylinder walls and surrounding atmosphere) forming a pressure above the liquid. This, as in the water experiment, has raised its boiling point, and when the pressure has reached the point where the corresponding boiling point is equal to the day temperature, evaporation will stop, since no further heat is available. Any reduction in the day temperature will cause some of the gas to condense, thus reducing the gas pressure. Any increase in temperature will restart boiling, raising the gas pressure.

For this reason gas cylinders must not be exposed to excessive temperatures, for example in the sun on a hot day, otherwise they may be subjected to a dangerous pressure.

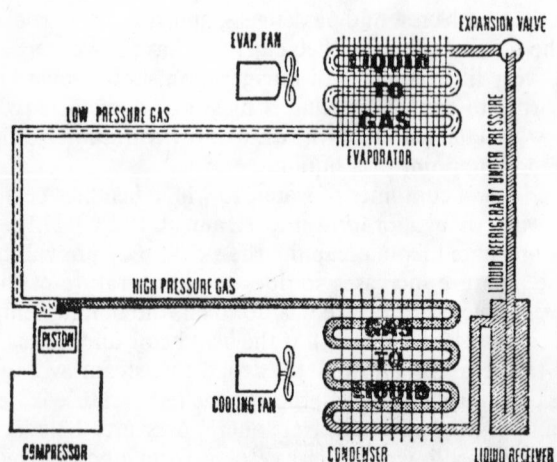

Fig. 13. Compression system of refrigeration, showing the continuous cycle of gas to liquid, and liquid to gas, taking place in a refrigerating system.

It should be clear, then, that if the cylinder valve is opened allowing the gas to be released, the pressure will be reduced, and the boiling point lowered. The day temperature will be sufficiently high to boil off the refrigerant, which it will continue to do all the time the gas is escaping. Heat continues to be taken through the cylinder walls, explaining why it becomes cold or even frosts if the action is allowed to continue for very long.

Mechanical cooling is now being achieved, but of course the escaping gas is being wasted, and unlimited supplies of refrigerant cannot be used in this extravagant way.

The refrigerator plant therefore is not so much a contrivance for extracting heat, as a method of providing a continuous flow of liquid refrigerant, using the same quantity of gas over and over again.

THE STAGES OF REFRIGERATION

Figure 13 illustrates a simple refrigeration system which may be divided into four stages as follows:

COMPRESSION
The compressor compresses the refrigerant, raising its pressure and temperature, and discharges it to the condenser.

LIQUEFACTION
Gas, cooled in the condenser whilst under pressure, is changed into a liquid, and discharged to the control valve.

EXPANSION
The control valve allows only a regulated amount of liquid through it, and the pressure beyond the valve is much lower.

EVAPORATION
Beyond the control valve is the evaporator, under a comparatively low pressure, and this sudden pressure drop causes the liquid passing into it to boil back into a gas, taking heat from the evaporator in the process. The gas is now drawn back to the compressor for the cycle to be repeated.

Putting this cycle into practical terms we have:

COMPRESSOR
The compressor is a mechanical device which draws in refrigerant gas, and discharges it at a higher pressure to the condenser. It is normally driven by an electric motor.

CONDENSER

The condenser is an apparatus which extracts sensible and latent heat from the pressurised gas, and transfers it to a cooling medium of air or water. The gas is thus liquefied.

LIQUID RECEIVER

A liquid receiver is a container which is situated immediately below the condenser to collect the liquefied refrigerant.

EVAPORATOR

An evaporator is an apparatus which absorbs heat from its surroundings, by evaporation of the liquid refrigerant metered into it. The liquid is thus transferred into a gas.

All mechanical refrigerator systems, no matter what the application, whether large or small, use this principle, and it is a useful exercise for the student to identify for himself the compressor, condenser, liquid receiver and evaporator, when attending various plants.

PUMPING DOWN

This is the term given to the process of shutting the valve on the outlet side of the liquid receiver so that the refrigerant within the system can be liquefied and trapped. A suction gauge must be fitted to the compressor, the plant started, and the gas reduced to a very slight gauge pressure. Normally the full refrigerant gas charge may be isolated in this way, enabling a repair to be carried out on the system.

Note: Do not open the system if the pressure has been reduced to below zero gauge, otherwise air will be allowed into the system.

If the 'pumping down' action has been carried on too long, resulting in a pressure below zero gauge, then (having stopped the compressor) 'crack' the liquid receiver valve briefly to raise the pressure again, or supplement from the service gas cylinder, until the required pressure is obtained. Then shut the compressor suction valve to the system (front-seated: *see* p. 47). Any part of the system between the liquid receiver valve and compressor suction valve may now be disconnected, and parts replaced as necessary.

6

Gauges, Vacuum, Heat Transfer

The two most important items in any refrigeration tool kit are the gauges. They are the eyes capable of penetrating the inner activities of a refrigerator system, and without which an engineer would be as helpless as a ship minus a compass.

Gauges are delicate instruments and the refrigeration apprentice must quickly learn to treat them with care and respect. On no account should they be permitted to mingle loosely in the tool-box, jostling with more robust tools. Unfortunately this is their usual fate, inevitably leading to scored or broken glasses, making, in the first place, visibility difficult, and in the second, grimy dials with risk of damage to the needle or mechanism. Information imparted by gauges in this condition is more of a menace than of assistance in diagnosing faults.

Accurate and reliable gauges are so necessary that a specially made box is recommended to protect them, on the lines of that illustrated in Fig. 14. This is constructed of wood, has a Perspex top as shown so that the dials may be read without being exposed, and should be long enough to accommodate the attached flexible lines—which of course must be plugged. Where a gauge manifold is used (obtainable from any refrigeration wholesaler) it can be incorporated with the handwheels outside the box on either side.

On examination of the two gauges, it is obvious that they are not identical. One dial shows readings from 0–300 lb/in^2, whilst the other only from 0–150 lb, with a minus or vacuum reading of 30 in on the other side of 0. A pair of SI gauges will probably read from 0–20 bar and 0–10 bar with a vacuum scale down to 760 mm. The first is to record the discharge pressure of a compressor, known as the pressure gauge, and the other the suction pressure, known of course as the suction gauge.

The zero reading on each dial is recording atmospheric pressure, and

27

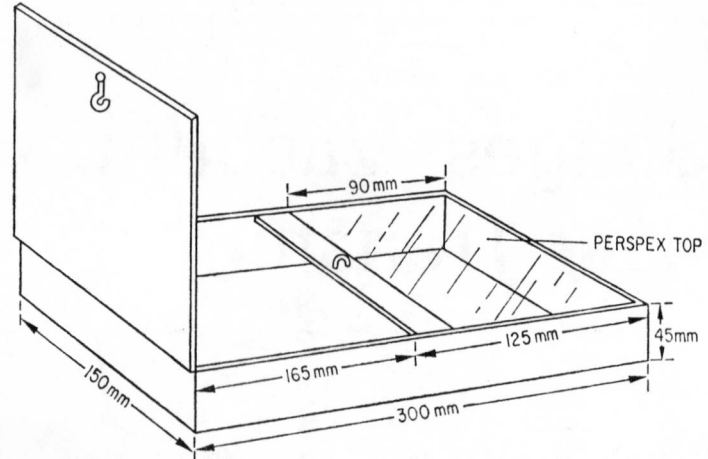

Fig. 14.   Box for gauges. Always take great care of your gauges. A special box
as shown is suggested.

is known as zero 'gauge pressure'. Some European gauges may be
marked in kg/cm². Except for very accurate measurements 1 kg/cm²
may be assumed to equal 1 bar. Atmospheric pressure was discussed
in the last chapter, and this seems an appropriate moment to
describe in greater detail what is meant by a vacuum, and why it is
measured in inches or millimetres.

A VACUUM AND HOW IT IS MEASURED

As already explained, the pressure exerted by the earth's atmosphere
at sea level is 1 bar (approximately 14·7 lb/in²). An experiment
to illustrate this may be carried out using a bowl of mercury (mercury
being the heaviest known liquid) and a glass tube, about one metre
long, and sealed at one end. If the tube is completely filled with
mercury and then the open end inverted into the bowl, some of
the liquid in the tube will drain into the bowl, until it drops to a
height of approximately 760 mm (30 in). At this level it will stop,
since the weight of mercury in the tube is being balanced by the
atmospheric pressure on the liquid in the bowl (*see* Fig. 15). It is
interesting to note that if water were used in this demonstration,
a tube more than 10 m high would be required to provide the same
balance.

ABSOLUTE PRESSURE
    The space above the mercury is now an almost complete vacuum.
For most practical purposes, therefore, every 2 in by which the

height of the mercury column is reduced represents an absolute pressure of 1 psi. In other words, absolute pressures start from a complete vacuum, gauge pressures start from atmospheric pressure.

Similarly, on an SI suction or compound gauge an absolute pressure of 0·1 bar would be shown as (760 − 76) = 684 mm.

Pressures greater than atmospheric may also be converted to absolute pressures by adding 15 lb/in² to an Imperial gauge reading or 1 bar to an SI gauge.

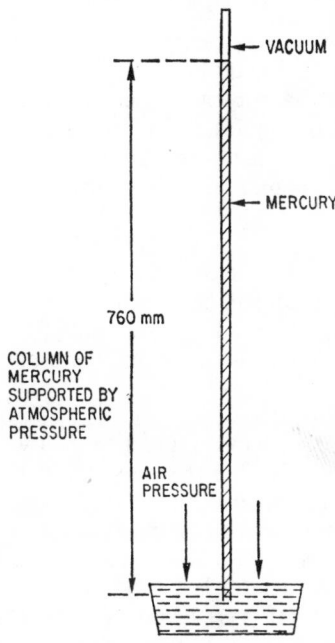

Fig. 15.   An experiment which illustrates what is meant by a vacuum.

Any pressure less than atmospheric is often referred to by refrigeration mechanics as a vacuum. The significance is that if the pressure in any part of a refrigeration system goes below atmospheric, then the greater outside pressure will tend to force air (and moisture) through any leak there may be in the system.

DEFINING A VACUUM

A *vacuum* may be defined as:

A pressure less than that of atmospheric pressure at sea level.

Absolute pressure may be defined as:

Pressure starting from a maximum vacuum.

Absolute pressure can be obtained by adding 15 lb to the registered gauge pressure.

HEAT TRANSFER

Heat is transferred naturally from one place to another in the following manner:

1. CONVECTION

If air (or liquid) is heated it expands. In so doing it becomes less dense and consequently lighter than the surrounding air. It therefore rises, to be replaced by cooler heavier air. This process creates a circulation of the air (or liquid), and is known as convection currents. The saying that heat rises is not strictly true, since it is the expanded air or liquid which is rising, taking the heat with it.

2. RADIATION

In a vacuum no such convection currents can be induced since no air exists, but heat can still be transmitted by means of radiation, and proof of this is provided by the heat of the sun reaching us through millions of space miles containing no atmosphere.

3. CONDUCTION

A third method is by means of direct contact with a hot substance against a cooler one, or within the substance itself. Evidence of this is given if one end of a poker is placed into a fire and allowed to get red hot. If now removed clear of the fire, the heat will gradually travel along the handle until it is too hot to hold.

THE SECOND LAW OF THERMODYNAMICS

This seems an opportune time to quote the *second law of thermodynamics* which states:

Heat is always transferred from a hot body to a cooler one.

In other words, heat always flows downhill and may be likened to a full tank of water trying to find its own level by running into a lesser filled tank, until both are equally filled.

This may at first appear to be an unimportant fact, and yet it is extremely useful when people talk of 'cold penetrating a room' or 'cold escaping through a coldroom floor'. Those conversant with the behaviour of heat will be able to explain that 'cold' does nothing, and that it is heat flowing away to a cooler level.

It is the object of a refrigeration system to reverse this process and actually 'pump' the heat 'uphill' out of the coldroom into the warmer surroundings.

AMBIENT TEMPERATURE

The word 'ambient' is bandied about very freely in refrigeration circles. It means 'surrounding temperature', and whilst it may mean more generally 'prevailing temperature', in practice it is applied more specially to the temperature surrounding the refrigerator unit. Thus the day temperature may be a moderate 21°C (70°F), but the 'ambient' or temperature surrounding the condensing unit 32°C (90°F) (or much more) and it is the ambient temperature, as the apprentice will soon discover, which is of most concern to the engineer.

PROGRESS CHECK

At the beginning I pointed out that this book would deal particularly with the practical side of an apprentice's training, and stressed that his ability to learn and to progress would depend very much on his own initiative in working out answers to problems himself. I cannot over-emphasise the importance of observation, and the asking of questions. It seems appropriate therefore to consider a few further points relating to day-to-day refrigeration service work, and to revise some of the facts already discussed.

It is common practice in many refrigeration service companies for an apprentice to be paired with one particular engineer most of the time. For this reason an apprentice often has time on his hands in the mornings whilst waiting for the engineer to hand in his service sheets, and this period can be put to good use in many ways.

For example, the van always benefits from a straightening up. When the day's programme is under way it so often happens that arriving at the next call takes priority over tidiness, but this state of affairs must not be allowed to get steadily worse. In addition, gas cylinders need to be weighed daily and replenished if required, and other items such as oil, paraffin, methylated spirits, all need to be regularly checked. An apprentice can be of great assistance by attending to these matters.

The outward appearance of the van is also worthy of attention, and at least once during the week time must be found to give it a wash down, and perhaps a polish.

If by chance the apprentice is waiting in the workshop there is bound to be ample scope for a sweep up. Also a general clearance of the work benches, bearing in mind of course not to disturb an unfinished repair, or to dispose of parts likely to belong to it.

An apprentice must develop a pride in his ability to be clean and tidy in all respects, for it is certainly true to state that efficient work cannot be carried out in a disorderly atmosphere, whether it be in the workshop, van or tool-box.

CARE OF TOOLS

Speaking of the tool-box, most apprentices will have acquired one at an early stage of their training, and made headway since in gathering together a selection of tools. I hope my earlier suggestion of counting them after each job has not been forgotten, and another tip is to have a piece of cloth in the box to give the tools a wipe over at the same time. Even with this procedure being conscientiously observed, it is advisable to empty the tool-box weekly, not only to give the contents a more thorough check, but to remove the accumulation of oddments which will make their home there if allowed to do so.

TEST QUESTIONS

Before proceeding further the student is advised to answer Test Questions No. 1 set on p. 139, at the end of the book. These are based on the facts covered in the preceding chapters, and if some appear simple, it will at least indicate that information is being absorbed. It is the simple facts which are so important in the early stages of any study.

In the course of a refrigeration service engineer's career, he will always need to be resourceful, for refrigeration must at all times be maintained. Thus on many occasions, simply out of necessity, he will have to improvise, and make do with parts which are not exactly suited to the job, but there is no substitute for knowledge.

7

Pressures and Temperatures

By observing the standard pressure gauges used in refrigeration, it will be seen that *temperatures* of common refrigerants are also given in relation to *pressures*. This important feature of service and installation work must be clearly understood, and the temperature/pressure relationship should always automatically be noted when reading the gauge so that the two become synonymous. In fact, it can be said that in most cases the gas temperature is more significant than the pressure.

Tables 7.1 and 7.2 show the temperature/pressure relationship for refrigerants R.12, R.22 and R.502, and are used to relate condensing temperature to the high side pressure, and evaporating temperature to the low side pressure.

Since charts are not convenient for carrying about, a plastic pocket 'comparator' is recommended, which readily provides the temperature/pressure comparisons, and every mechanic must have one of these with him at all times.

The pressure/temperature figures provide vital information based on the facts summarised as follows:

a. If the pressure is maintained, any cooling will result in the gas being liquefied.
b. Any heat available above the temperature/pressure obtaining will result in further evaporation of the liquid, and a consequent increase in pressure.
c. Any reduction of pressure below the temperature/pressure obtaining will permit the liquid to evaporate, taking heat from its surroundings.

This process has already been illustrated in connection with the gas cylinder in Chapter 5, but Tables 7.1 and 7.2 detail the pressures and temperatures with which we are concerned. Thus with a day

TABLE 7.1
Pressure/Temperature Relationships (Imperial Units)

Temp. °F	R.12	lb/in² gauge R.22	R.502	Temp. °C
−40	11·0*	0·5	4·3	−40·0
−35	8·4*	2·6	6·7	−37·2
−30	5·5*	4·9	9·4	−34·4
−25	2·3*	7·4	12·3	−31·7
−20	0·6	10·1	15·5	−28·9
−15	2·4	13·2	19·9	−26·1
−10	4·5	16·5	22·7	−23·3
−5	6·7	20·0	26·8	−20·6
0	9·1	24·0	31·2	−17·8
5	11·8	28·2	36·0	−15·0
10	14·6	32·8	41·0	−12·2
15	17·7	37·7	46·6	−9·4
20	21·0	43·0	52·4	−6·7
25	24·6	48·7	58·7	−3·9
30	28·5	54·9	65·4	−1·1
35	32·6	61·5	72·6	1·7
40	37·0	68·5	80·2	4·4
45	41·7	76·0	88·3	7·2
50	46·7	84·0	96·9	10·0
55	52·0	92·5	106·0	12·8
60	57·7	101·6	115·6	15·6
65	63·8	111·2	125·8	18·3
70	70·2	121·4	136·6	21·1
75	77·0	132·2	148·0	23·9
80	84·2	143·6	159·9	26·7
85	91·8	155·7	172·5	29·4
90	99·8	168·4	185·8	32·2
95	108·3	181·8	199·6	35·0
100	117·2	195·9	214·4	37·8
105	126·5	210·7	229·7	40·6
110	136·4	226·3	245·8	43·3
115	146·8	242·7	262·6	46·1
120	157·6	259·9	280·3	48·9

* Inches of mercury

TABLE 7.2

Pressure/Temperature Relationships (SI units)

Temp. °C	Absolute pressure bar (For gauge pressure bar, deduct 1)		
	R.12	R.22	R.502
−40	0·64	1·05	1·31
−39	0·67	1·10	1·37
−36	0·77	1·26	1·56
−33	0·88	1·44	1·77
−30	1·004	1·63	2·00
−27	1·14	1·85	2·25
−24	1·29	2·09	2·53
−21	1·45	2·35	2·83
−18	1·63	2·64	3·16
−15	1·83	2·95	3·51
−12	2·04	3·29	3·90
−9	2·27	3·66	4·32
−6	2·52	4·06	4·76
−3	2·79	4·49	5·25
0	3·09	4·96	5·76
3	3·40	5·47	6·32
6	3·74	6·00	6·91
9	4·11	6·59	7·54
12	4·50	7·21	8·21
15	4·91	7·88	8·93
18	5·36	8·59	9·69
21	5·83	9·34	10·50
24	6·34	10·15	11·35
27	6·88	11·00	12·26
30	7·45	11·92	13·22
33	8·05	12·89	14·23
36	8·69	13·91	15·30
39	9·37	14·99	16·42
42	10·09	16·13	17·61
45	10·84	17·34	18·86
48	11·64	18·61	20·17
50	12·19	19·50	21·09

Acknowledgement is made to E. I. Du Pont de Nemours and Co. for permission to extract figures for R.12, R.22 and R.502 from tables issued by them and to I.C.I. Ltd. for permission to use information from their pressure/temperature chart in SI units.

temperature of 30°C, the pressure in the R.12 cylinder would be 6·45 bar, and in the R.22 cylinder it would be 10·92 bar. A similar relationship can be obtained from the Imperial chart.

TEMPERATURE DIFFERENCE

The transfer of gas to liquid, and liquid to gas, takes place continuously in a refrigeration system, but before enlarging on this, an explanation of temperature difference (TD) seems appropriate.

To cool a substance it is necessary for the cooling medium to be lower in temperature than the substance. The second law of thermodynamics (p. 30) explains that heat flows from a hot body to a cooler one, and the two cooling mediums readily available to reduce the gas temperature in refrigerator units are air and water. Since these must be used at their prevailing temperatures, to obtain a temperature difference (TD) the gas temperature must be raised above the temperature of the cooling medium. Only when this is achieved will it be possible to cool and thus liquefy the refrigerant.

For example, if cooling water is 15°C the gas temperature must be raised to, say, 24°C, which from Table 7.2 for R.12 is seen to be a pressure of 5·34 bar.

In Imperial units, assuming that the cooling water is 60°F the gas temperature must be raised to, say, 75°F, which from Table 7.1 is seen to be 77 lb. The same relationship can be obtained from the tables for R.22 and R.502.

Let us consider for a moment a standard coldroom plant and see how the temperature/pressure tables relate in practice. The following are the relevant facts:

	SI	Imperial
Temperature required in the coldroom	3°C	35°F
Ambient temperature	30°C	85°F
Refrigerant	R.12	R.12
Condenser air cooled in both cases		

Taking the ambient temperature, which is the cooling medium in these cases, and adding a TD in SI of 9°C and in Imperial of 15°F, we have:

SI	Imperial
30 + 9 = 39°C	85 + 15 = 100°F
Pressure for R.12 = 8·37 bar	Pressure for R.12 = 117·2 lb

These are the pressures expected to be shown on the discharge gauges. Owing to a relatively high suction pressure (and remembering that the tables are a guide only) they will be higher than these, perhaps 9·84 bar or 130 lb.

Inside the coldroom, to maintain a temperature of 3°C or 35°F the evaporator must be lower in temperature, and again we apply a 9°C TD for SI and 15°F for Imperial units, thus:

SI	Imperial
$3 - 9 = -6°C$	$35 - 15 = 20°F$
Pressure for R.12 at	Pressure for R.12 at
$-6°C = 1·52$ bar	$20°F = 21$ lb

These are the pressures to be expected on the suction gauges.

The suction pressure—or 'back pressure' as it is usually called—is seen to be of particular importance, since it must be regulated or adjusted to suit the required temperature. It would be of no use to evaporate at the pressure of 1·52 bar, for example, which we have seen is $-6°C$ if the required temperature was, say, $-12°C$; or, in Imperial, to evaporate at 21 lb if the required temperature was 10°F. In these cases we have different pressures:

$$-12°C - 9°C = -21°C \quad \text{or} \quad 10°F - 15°F = -5°F$$

To obtain these temperatures we need 'back pressures' of:

$$0·45 \text{ bar} \qquad 6·7 \text{ lb}$$

There are several methods of controlling the back pressure which, by the way, is not necessarily constant, and these will be discussed later. At this stage of progress the main concern is to explain the reasons for certain pressures, for only by knowing what the pressures should be, is it possible to judge if they are incorrect.

In present-day refrigeration installations, several temperature ranges are in general use, and Table 7.3 lists standard operating temperatures for different applications, together with normal back pressures to be expected on plants using R.12 refrigerant. The student may like to check these from Tables 7.1 and 7.2 and also to see what the pressures would be if R.22 were being used.

The 'head' or discharge pressure will not vary a great deal, although, as already stated, a high back pressure will result in a higher head pressure. It is perhaps advisable to point out that where a condensing unit is operating normally with its gas charge correct, and condenser being adequately cooled, no adjustment of head pressure is possible or, of course, necessary.

Since the compressor is running at a constant speed, the back pressure is regulated by increasing or decreasing the flow of refrigerant. Decreasing the flow must also result in reducing the quantity of heat being extracted, but this sacrifice has to be made in order to obtain a lower back pressure at a lower temperature. Maximum

TABLE 7.3

Standard Operating Temperatures and Expected Back Pressures on Plants Using R.12 Refrigerant

Application (R.12 refrigerant)	Temp. required		Evaporation temp.		Back pressure gauge	
	SI °C	Imp °F	SI °C	Imp °F	SI bar	Imp lb/in²
Air conditioning	20	68	11	53	3·36	50
Cellar cooling	13	55	4	40	2·51	37
Dairy coldroom	5	40	−4	25	1·70	25
Meat coldroom	2	35	−7	20	1·44	21
Fish coldroom	0	32	−9	17	1·27	19
Meat coldroom	−2	28	−11	13	1·11	16
Meat coldroom	−7	20	−16	5	0·76	12
Frozen food cabinet	−18	0	−27	−15	0·14	2
Hardening room	−23	−10	−32	−25	60 mm Hg	2 in vacuum

efficiency is effected, therefore, by maintaining as high a back pressure as possible, consistent with maintaining the required evaporating temperature.

8

Refrigerant Flow Control

The pressure generated in a refrigerator compressor, apart from creating the conditions necessary to liquefy the refrigerant, also serves to provide a means of propelling the liquid to the evaporator. It has already been shown (Fig. 13) that liquid is formed in the condenser from gas being cooled whilst under pressure, and this liquid passes via the liquid receiver to the evaporator, still under pressure. The pressure forces liquid beyond the receiver to the control valve, and so into the evaporator. There the liquid evaporates and the resultant gas is drawn back to the compressor. This cycle is, no doubt, a familiar one to most apprentices, but it is repeated to emphasise the importance of the control valve.

The whole system is regulated at this point, the function of the valve being to hold up the refrigerant flow and to meter it through at the rate required to produce the maximum quantity of refrigeration at the desired temperature. The last chapter described the reasons for certain pressures, and the purpose of this one is to discuss the various methods by which they are achieved.

Any ordinary shut-off valve would be adequate if it were opened just sufficiently to produce the correct back pressure, and regulated from time to time to maintain it. In some old installations this method is in fact used, but as it entails constant attention, it is of no use to the smaller units which must be automatic in operation. The most common methods of refrigerant flow control are listed as follows.

CAPILLARY TUBE

This is not a valve, merely a long fine-bore copper tube, usually about 1 mm (0·03 in) bore, and between 2 and 4 m (8 and 12 ft) long.

The flow of refrigerant is restricted by the tube, and the correct 'back pressure' is obtained in the first instance by adjusting its length. Once this has been established, the tube is brazed into position and no further adjustment is possible. The capillary system is used universally on domestic refrigerators, and widely used on small ice cream or frozen food cabinets. It is a cheap and simple means of controlling refrigerant flow, with no moving parts to go wrong, and furthermore it is readily adaptable to a small gas charge which will balance to an equal head and back pressure during the 'off' cycle, thus reducing the starting torque on the compressor motor (*see* p. 82).

FLOAT

This method of control will not be found very often these days, except in the large industrial ammonia plants. It is used in what is known as a flooded evaporator, so called because it contains a liquid refrigerant, which, as it boils off, is supplemented via a float and valve, similar to a ball valve on a water tank. This method involves relatively large quantities of refrigerant, and as a result the evaporators are somewhat cumbersome. Many pre-war domestic refrigerators employed float control, but most of them have now become obsolete.

AEV (AUTOMATIC EXPANSION VALVE)

This valve is designed to maintain a constant pressure in the evaporator, and is illustrated in Fig. 16. The pressure at point A is equal to the pressure in the evaporator, and as the compressor extracts gas from the evaporator, the pressure at point A is reduced. This allows the range spring to overcome the balance spring (via the push rod) thus opening the needle valve. If the pressure at point A rises again it will press up against the range spring, so allowing the needle valve to close down. In this way a balance is reached, and since the compressor is running at a constant speed, the evaporator pressure remains steady. By adjusting the tension on the range spring, using the adjusting screw, the constant pressure may be raised or lowered as required.

The 'back pressure' therefore remains constant at the pre-set pressure. This is an advantage in that the evaporator temperature and the load on the compressor remain constant, and at the same time a disadvantage, in that the plant is incapable of adapting itself to fluctuating loads. For example, when a butcher takes in an

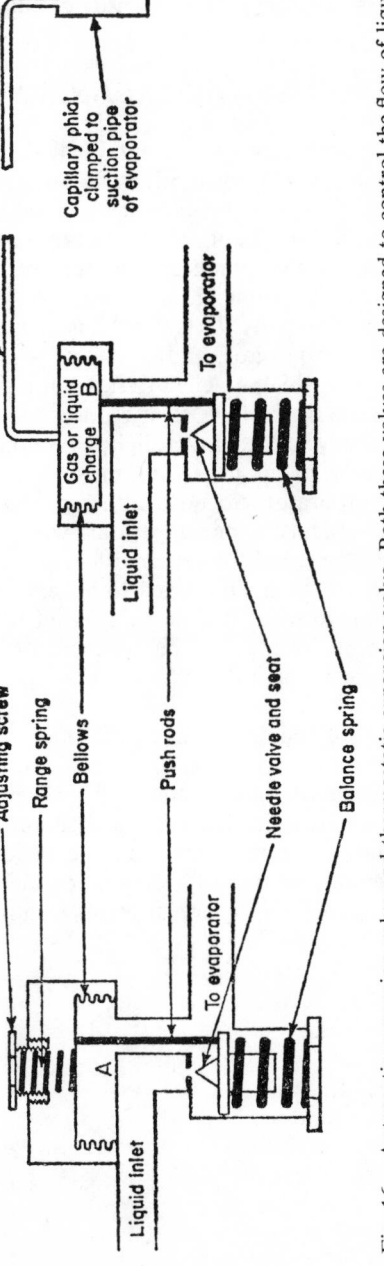

Fig. 16. Automatic expansion valve and thermostatic expansion valve. Both these valves are designed to control the flow of liquid refrigerant in the refrigeration system.

extra amount of meat into his coldroom, an AEV cannot allow
more liquid to pass to increase the quantity of cooling.

TEV (THERMOSTATIC EXPANSION VALVE)

As a result of the limitations just mentioned, the TEV (illustrated
in Fig. 16) was developed. Basically it is an AEV to which the
refinement of a capillary and bellows has been added. (The bellows
is described more fully in Chapter 12.) The capillary tube is charged
with a gas or liquid which expands or contracts readily with temperature changes, and this action automatically performs the duty of
the adjusting screw of an AEV. The phial is clamped firmly to the
suction pipe of the evaporator, and as the evaporator is cooled, so
gradually is the valve phial attached to it, resulting in a contraction
of the medium inside the bellows. This reduces the pressure at point
B, and the needle valve closes down slightly (via the push-rod), thus
reducing the quantity of liquid passing through it and lowering the
pressure in the evaporator. So now, although the amount of heat
extraction has been reduced, the temperature will have been lowered.
This process continues until the required temperature is reached.

Should a butcher have a TEV fitted to his coldroom evaporator,
on increasing the load, heat will be transferred to the phial of the
valve, raising the pressure in the bellows and opening the needle,
thus allowing more liquid through and so adapting the supply to
the demand.

These are the most common forms of refrigerant flow control,
the TEV being the most widely used. Manufacturers supply ample
literature illustrating the construction of their products, together
with instruction on adjustments, and these should be studied carefully. The smallness of the orifice and the delicacy of the TEV
makes it a vulnerable source of refrigeration failure—not always
the fault of the valve—and every apprentice must understand its
operation fully.

9
Refrigerants

So far, although the action of a refrigerant in a system has been described, and the way in which it is liquefied and evaporated explained, the reason why particular gases are used has not been discussed.

To the refrigeration apprentice, 'gas' quickly becomes an everyday word, and the 'gas' cylinder a familiar object. It is to be hoped that the enquiring student will have asked why it is that certain gases are used as refrigerants. There is no special magic about them, many alternatives could be used, but those chosen do possess a reasonable balance of the practical and thermodynamic requirements of a good refrigerant.

No two substances have the same characteristics, as reference to the tables in Chapter 7 will clearly illustrate, and in the selection of a suitable gas the following are the main factors which have to be considered:

1. The pressure required to liquefy the gas in the condenser, with a cooling medium of air and water at, say, 20–30°C (70–80°F).
2. The pressure required in the evaporator to achieve the necessary temperature of evaporation.
3. The quantity of heat which a given mass of refrigerant will extract during evaporation.
4. The density of the vapour.
5. Whether inflammable or toxic.
6. Whether a leak is easily detectable.
7. Whether harmful to commodities.
8. Whether it has a corrosive action on metals in the machinery.
9. Price.

Taking these points one at a time, the pressure in the condenser is, of course, produced by the compressor, and a very high pressure causes extra stresses on the piston, cylinder head, discharge piping, and to the condenser itself.

The same condition applies to the evaporator, which must be robust if, in this case, the suction pressure of the compressor is high.

CARBON DIOXIDE, OR AMMONIA

It is of interest to compare carbon dioxide (CO_2) and ammonia (NH_3). Neither of these refrigerants will be found in the small automatic machines in common use, but their properties do serve to illustrate the differences to be found in two refrigerants.

On the question of pressures CO_2 is at a distinct disadvantage, for whereas with NH_3, using cooling water at, say 18°C (65°F), the head pressure would be about 10 bar (150 lb/in²), that of CO_2 would be nearer 60 bar (900 lb/in²). On the suction side, assuming the liquid to be evaporating at -12°C (10°F), the back pressure for ammonia would be 1·6 bar (24 lb), but for CO_2 24 bar (360 lb).

To those familiar only with more modern refrigerants, a back pressure of close on 27 bar (400 lb) must seem incredible.

Prime considerations in a refrigerant, therefore, are a low condensing pressure, together with a low back pressure (although not so low as to produce a vacuum) and the differential between the two pressures to be as little as possible.

Turning now to the refrigerant in liquid form, to evaporate any liquid a given quantity of heat is required, depending on the substance, and the greater this amount is, the more refrigeration will be achieved. The refrigeration effect which can be obtained by evaporating a given quantity of ammonia is more than seven times as much as can be obtained with the same quantity of CO_2.

At first sight, therefore, CO_2 would appear to be an inefficient refrigerant, but there is more to it than that. When evaporated, the volume of ammonia vapour produced will be more than thirty times as great. Thus with the same compressor speed and displacement, CO_2 can be circulated round a system thirty times as often as ammonia, so in actual fact CO_2, taken on this basis alone, is a more effective refrigerant. Also it allows for a more compact unit, and size of machinery must always be an important factor when considering the relative merits of a refrigerant.

The question of safety is another important feature, for the potential danger from large quantities of explosive gas escaping is obvious, and in some cases safety is the deciding factor in the choice of a refrigerant. Ammonia is inflammable, CO_2 is not, and in the case of ship refrigeration, for example, a non-inflammable gas has always been preferred. On the other hand, ammonia is easily detectable owing to its smell, whereas CO_2 leaks are not apparent in this way. Its high pressure, however, makes it readily detectable by sound, or by a soapy solution brushed on to the joints.

Ammonia has a corrosive action on copper, thus steel evaporators are necessary, introducing another problem because steel will rust, and the exterior piping, being exposed to frost and moisture, has

TABLE 9.1
Properties of Common Refrigerants

	Imperial units	Condensing temperature 30°C (86°F) Evaporating temperature −15°C (5°F)			SI units			
		R.12 (CCl_2F_2)	R.22 ($CHClF_2$)	R.502 ($CHClF_2$ 48·8% + $CClF_2CF_3$ 51·2%)		R.12	R.22	R.502
Head pressure	lb/in²	93	158	177	bar/g	6·45	10·9	12·25
Suction pressure	lb/in²	12	28	36	bar/g	0·83	1·93	2·49
Refrigerating effect	Btu/lb	68	93	69	kJ/kg	159	217	161·4
Specific volume	lb/ft³	1·46	1·24	0·825	m³/kg	0·091	0·077	0·051
Inflammable or toxic?		No	No	No		No	No	No
Leaks readily detectable?		Yes	Yes	Yes		Yes	Yes	Yes
Harmful to commodities?		No	No	No		No	No	No
Corrosive to metals?		No	No	No		No	No	No

to be galvanised. Steel piping is invariably used with CO_2, not because of any corrosive action, but because of the strength needed to withstand the pressure.

Two other gases which should be mentioned in this survey of refrigerants are sulphur dioxide (SO_2), now almost obsolete as such, owing to its toxic quality and low back pressure, and methyl chloride (CH_3Cl). Methyl chloride will still be encountered in early post-war plants, but has now been abandoned mainly owing to its inflammability. It must on no account be put into a sealed-unit system owing to its action on the internal metals, but it is permissible to change a methyl chloride plant to R.12, provided, of course, that the system is thoroughly pumped out first, and the expansion valve adjusted or changed.

By far the most frequently used refrigerants today are R.12 and R.22, with R.502 becoming increasingly used, and the refrigeration apprentice will find it a useful exercise to study the facts set out in Table 9.1, listed in the order given before.

From the details given it can be seen how these gases, produced over recent years, satisfy most of the demands of an ideal refrigerant.

10
The Compressor

The heart of a refrigeration system is the compressor. If this is not functioning correctly, then any other adjustments made or parts replaced can give only partial improvement.

A refrigeration engineer in many ways resembles a doctor, in that he is asked to attend a plant when it is not up to the mark, and in the same way, as a preliminary to diagnosing the trouble, produces a thermometer to check the temperature. He will then ask the owner to describe a few symptoms, meanwhile feeling the suction and discharge pipes for temperature, akin to taking a pulse, and then, like a doctor with a stethoscope, he puts on his gauges to find out what is taking place inside.

Just as a weak heart will produce side effects in a human being, and knowledge of its condition is vital to diagnosing a patient's illness, so it is wise for the engineer always to check the refrigerator compressor first, and to record its performance for future reference.

SERVICE VALVES

Before proceeding to the compressor itself, it is expedient to discuss the service valves, the special feature of which is their dual purpose. When shut forward (front seated), that is, the spindle fully turned in a clockwise direction, the compressor is isolated from the system, and when fully opened in an anti-clockwise direction (back seated) gas is isolated from the gauge port. This allows the plug to be removed, and the gauge connections fitted, without losing refrigerant. A study of Fig. 17 is advised to understand fully the terms mentioned of 'back seated' and 'front seated'.

During normal operation the valve is 'back seated', which also prevents gas from leaking along the valve stem and through the gland. The valve has a cap protecting the spindle end, which is squared to take the valve key, and this cap is tightened up on to a copper ring as shown, which also assists in preventing a gas leak

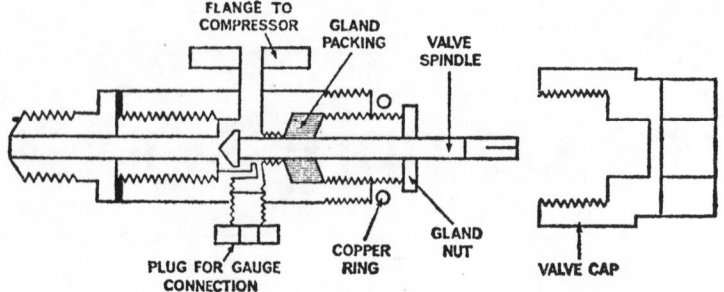

Fig. 17. Compressor service valve. This valve shuts to the system when 'front
seated', and to the gauge port when 'back seated'.

through the gland. This factor is of particular value when a low
pressure control is necessary. Unless a connection is specifically
provided on the compressor, the usual practice is to fit a 'T' in
place of the blanking plug as shown in Fig. 18.

Two connections are now available, one being used as a permanent
connection to the control, and the other for fitting a gauge connection
when required. When not required for this purpose, the connection
must be blanked off, and the usual means is to use an ordinary flare
nut, with a copper 'bonnet' inserted, as illustrated. The important
point in this case is first of all to 'back seat' the valve before connecting the gauge line, and then to crack the valve forward afterwards
about half a turn, allowing gas pressure to the control. A common
error is to 'back seat' the valve, the result of which is to eliminate
the function of the control.

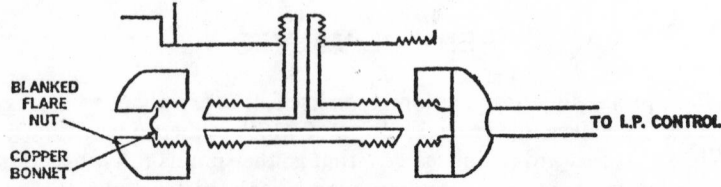

Fig. 18. 'T' piece fitted into service valve. After removing the plug as shown in
Fig. 17, the 'T' connection enables a gauge line to be fitted, and a permanent
line to a LP control.

PUMPING ACTION

The pumping action of a compressor is quite simple and straight-
forward (*see* Fig. 19). As the piston is pulled down in the cylinder,
gas is drawn in through the suction service valve, into the suction
manifold, and through the suction reed which flaps clear of its

seat during this stroke. The head reed meanwhile remains firmly seated by virtue of the pressure above it. As the piston is pushed upwards, the suction reed springs back on to its seat, thus preventing the gas from returning the way it came. The pressure now rises rapidly in the cylinder forcing the discharge reed to lift from its seat, and the compressed gas to pass into the discharge manifold, and to the discharge service valve. Most compressors have two cylinders, so that when one is drawing in gas, the other is discharging, thus providing steadier discharge pressure, and also a more even load on the driving motor.

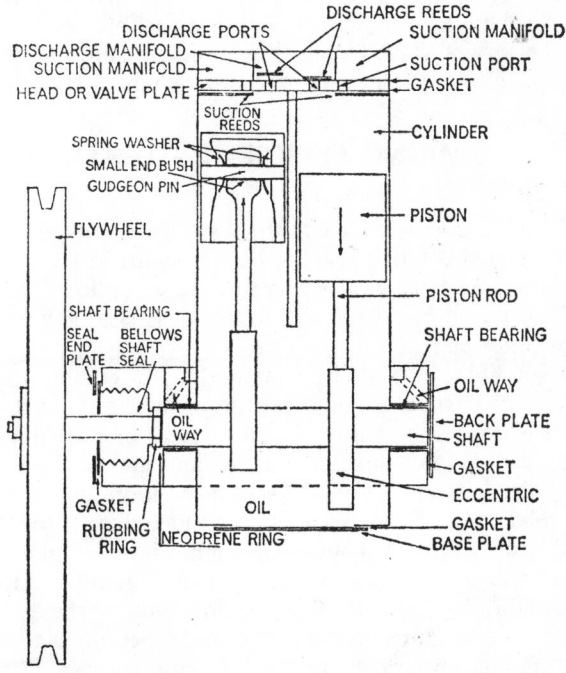

Fig. 19. A typical twin cylinder compressor. This illustrates the component parts of a compressor as used in the low horse-power condensing unit range.

MECHANICAL FAULTS

Inefficient pumping of a compressor may be due to one or all of the following:

POINTS OF FAILURE

1. Wear on the cylinder walls and pistons, causing gas pressure to leak into the sump.

2. Worn or broken piston rings, if fitted.
3. Leaking or broken valve reeds.
4. Too thick a cylinder head gasket, giving excessive clearance between the piston and valve plate.

LEAKING REEDS

By far the most common fault is leaking reeds, and since checking for pumping efficiency is such a simple matter, it is always advisable to do this as a first test. Even though the compressor may not be the prime cause of a fault, knowledge of its performance is useful information. After fitting gauges, make particularly sure there are no leaks on the lines. Do this by allowing gas into them and then shutting the service valve to the gauge and watching to see if the pressure remains. The following procedure is now carried out:

PUMPING EFFICIENCY TESTS

TEST 1

Shut the suction service valve to the system. Record the maximum vacuum obtainable, noting also the head pressure at the same time.

Reasonable time must be given for this test, since liquid refrigerant is sometimes mixed in the oil, and until this is cleared a maximum vacuum will not be reached.

This possibility can be confirmed by feeling the compressor casing at oil level, which will cool and even show evidence of condensation, and if this is happening ten or fifteen minutes or even longer may be necessary, but if not three or four minutes are usually sufficient.

As a guide, a 500 mm (20 in) vacuum or more, against a head pressure of 7 bar (100 lb) would be considered very satisfactory. Below 500 mm (20 in) but above 250 mm (10 in) would indicate some inefficiency, but a performance still good enough to maintain refrigeration. Below 250 mm (10 in) would warn that the compressor might be the main source of the complaint, especially if this was of 'temperature not low enough' or 'running continuously'.

When a reed is broken no vacuum at all is obtainable, and the gauge needle flaps up and down violently.

One point to watch in this test is the possibility of 'slugging' oil (oil in the sump being discharged through the reeds) which sometimes takes place if the sump is subjected to a low vacuum with an overcharge of oil. If this happens, and it is clearly evident by sound and vibration, stop the plant at once, otherwise damage will be caused to the reeds and to the head gauge. The procedure for this problem will be dealt with more fully later on, but for the moment let it be assumed the readings have been obtained without this happening.

TEST 2

With the compressor still running, and the maximum vacuum having been reached, shut the discharge service valve to the system and stop the plant. Watch both gauges and note how quickly the head pressure drops, and the suction pressure rises.

If the reeds are seating perfectly the pressure will remain steady, but if they are leaking badly, rapid equalisation will take place. The speed of equalisation is an indication of the efficiency of the head reeds.

Some compressors have a small 'breather' between the suction and head, in which case test 2 is inconclusive, but test 1 is the most important and provides most of the information required. The latter test can also be applied to semi-hermetic units, where a discharge valve is not normally fitted to the compressor itself, so test 2 is not possible anyway. Simply shutting the suction valve to the system, and noting the vacuum, generally takes only a matter of seconds, but the information it provides may save hours.

VALVE PLATE

Figure 20 illustrates the underside of a typical valve plate, showing suction ports and reeds, whilst Fig. 21 illustrates the upper side of the same valve plate, showing the discharge ports and reeds. The reeds are made of thin hardened steel, each head reed being held on to its seat by two springs, whilst each suction reed relies upon the spring action of the reed itself. In each case seating is assisted by the gas pressure, as already mentioned.

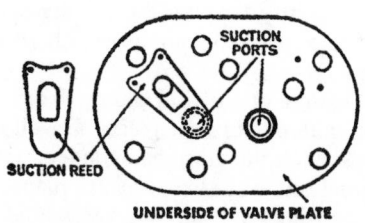

Fig. 20.   Underside of valve plate (from compressor in Fig. 19), showing the suction ports and reeds.

By examining a new head plate it can be seen how carefully the surfaces have been ground and polished, including of course the valve seatings, and it is this feature which ensures uniform contact with the reed, essential to provide efficient sealing. Any unevenness of the seating of the reed, and scoring of the seating, or any foreign

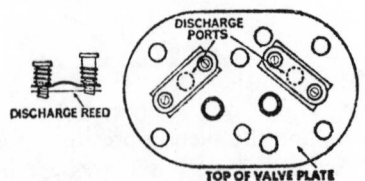

Fig. 21.   Upper side of valve plate. This is the same valve plate as in Fig. 20,
showing the discharge ports and reeds.

matter preventing the reed from seating properly, will adversely
affect the pumping efficiency. Deterioration of efficiency generally
takes place gradually, but if a compressor stops pumping suddenly,
it is usually due to dirt or a broken reed, and if faced with this
problem, an engineer must do something quickly to restore refrigeration. The normal method is to fit a new valve plate, complete with
reeds and gaskets, in the following manner:

FITTING A NEW VALVE PLATE

1. Shut the suction service valve to the system, the machine being
   stopped.
2. Shut the discharge service valve to the system, and release gas
   pressure in the compressor through the head gauge line.
3. Thoroughly clean compressor head and have clean rag available.
4. Remove head bolts, putting them into a safe place, preferably
   in a box.
5. Lift cylinder head slightly and remove valve plate. If stuck,
   ease apart by gently tapping the cutting edge of an old table
   knife into the gasket. Do not use a chisel as this may cause
   damage.
6. Clean off any remains of old gasket from cylinder head and
   compressor, making sure no pieces are allowed to drop into
   the compressor by plugging the cylinders with the clean rag.
7. Smear a film of oil on the new gaskets, place them into position,
   together with the new suction reeds and valve plate, and
   reassemble. Tighten bolts evenly with a well-fitting ring or box
   spanner.
8. Open suction valve slightly, remove head gauge and allow gas
   in the system to purge through the compressor to the atmosphere, giving the plant a run for a few seconds to assist
   clearance of air from the cylinders. Replace head gauge, and
   test compressor for pumping. If satisfactory, open up first the
   head valve and then the suction valve.

RENOVATING AN INEFFICIENT VALVE PLATE

There are occasions when a new head plate is not immediately available, in which case it is necessary to carry out first aid methods on site. These are similar to the procedure which would be carried out in the workshop, making use of the 'extras' list on p. 6.

Carry out procedure 1 to 6 as above, and then as follows:

a. Thoroughly clean off remains of old gasket from the valve plate. Remove discharge reed assemblies, putting springs etc. into a safe place.

b. Ensure glass is perfectly clean, and place the medium 'wet and dry' (carborundum paper) on to it with the grinding surface upwards. Place valve plate on to the 'wet and dry' and, pressing fairly heavily, begin to move it back and forth, changing the position of the plate from time to time. It is important that the paper should remain flat during this process, and not be allowed to crinkle. Continue in this way, changing the paper if necessary, until it can be seen that the seat surfaces have been affected. At this stage shorten the motion and give a figure 8 rotation. (If the seats are very bad a coarse paper may be necessary to start with.)

c. Change now to a fine 'wet and dry', and reduce the pressure, changing the rotation more frequently.

d. Finally polish by using a clean piece of cloth impregnated with metal polish, in place of the 'wet and dry'. Finish off by cleaning the valve plate in carbon tetrachloride.

e. From now on it is essential that hands are perfectly clean. Remove any burrs from the reeds and check for flatness by laying them on a clear part of the head plate and pressing on to one edge as in Fig. 22. Put a spot of oil on the seats, and refit the reeds, facing them on to the seats so that any 'bowing' will assist seating. Reassemble head springs. Check that head reeds are free to move up and down. Complete the repair by continuing with operations numbers 7 and 8.

When a head plate is badly scored, it has to be reground and honed by machine, but in some cases it is more economical to fit a new plate. However the method described above will certainly provide improvement as a temporary measure.

Fig. 22. Checking a reed. The 'bend' of a reed can be ascertained by placing it on the valve plate, and pressing one end as shown.

COMPRESSOR SHAFT SEAL

The most common cause of failure on a refrigeration system is loss of refrigerant, and the most vulnerable point at which a leak can develop is on the compressor shaft seal. It being impracticable to construct a compressor with the refrigerant isolated from the crankcase, and also impossible to drive a shaft unless it protrudes through the casing, the problem exists of designing a compressor which will allow the shaft to rotate freely and at the same time retain the gas pressure within. The gas pressure in the crankcase may well rise to 3–4 bar (50 or 60 lb), or even more, during the 'off-cycle', and it is necessary to emphasise that the shaft seal must be completely gas-tight under all stresses placed upon it during the normal operation of a condensing unit. The shaft seal is therefore a very important component.

BELLOWS SEAL

Figure 23 (*see also* Chapter 11) illustrates a standard type of bellows shaft seal, consisting of four parts as shown. When placed into its working position, the bellows portion is compressed so that

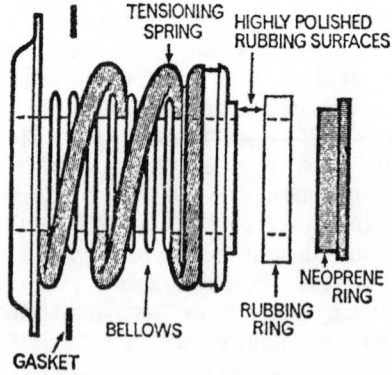

Fig. 23.   Bellows shaft seal. Refer to Fig. 19 for position of seal, which shows how gas in the compressor is prevented from escaping along the flywheel shaft.

a firm pressure exists between the polished surfaces of the bellows and the rubbing ring. This, aided by a continual mist of oil from the sump, achieves the result required. The ring so marked is made of Neoprene, a synthetic substance, very similar in texture and appearance to rubber, but which is impervious to oil. The purpose of this ring is to prevent gas seeping along the compressor shaft, and Fig. 19 shows the complete component in position.

A shaft seal leak is often difficult to locate, because it has a tendency to be intermittent, so when testing for leaks with a detector, it is advisable to move the flywheel to several positions, and also to examine the seal end plate for traces of oil. Oil will frequently betray a gas leak which might otherwise remain undetected, but it would be wrong to change a shaft seal solely on the evidence of a small quantity of oil, since some seals do weep oil without allowing a serious gas leak. This is particularly applicable to large compressors.

CHANGING THE SHAFT SEAL

Assuming for a moment that a compressor is accessible, or clear of the unit base, the procedure for changing the shaft seal is as follows:

1. Remove flywheel by means of suitable removers (*see* p. 57). Remove seal end plate, old gasket, and seal assembly.
2. Thoroughly clean seal compartment, removing all traces of the old seal, and polishing the shaft with fine 'wet and dry' carborundum paper, right up to the shoulder of the shaft. Remove any traces of burrs from the flywheel end, particularly on any key-ways.
3. Now ensure that hands are perfectly clean before unwrapping the new seal. Avoid touching the rubbing surfaces. Lightly oil the compressor shaft, and press the rubbing ring together with the Neoprene ring firmly up to the shaft shoulder, making sure that the highly polished side of the rubbing ring is facing towards the bellows.
4. Inspect rubbing surfaces once more, lightly oil, smear oil on the gasket. Place gasket and bellows into position, fit seal end plate, and insert all bolts about one turn.
5. Tighten bolts evenly, keeping the end plate parallel with the compressor flange, and turn the main shaft a time or two as the bellows is pressed in, to avoid the possibility of the eccentric binding on the connecting rods.
6. Before finally tightening up, ensure that the bellows is central to the shaft, if not move it into position, then tighten all bolts firmly, and replace flywheel.

CAUSES OF SEAL LEAK

CONTAMINATED OIL
These are the main mechanics of seal changing, but it is also wise to consider the possible cause of the seal leak in the first place.

In many cases this is due to normal wear and tear, in others to lack of oil, or to oil which has become contaminated.

Remedy

Where possible, therefore, it is advisable to remove the base plate, when removing the old seal, and to wash out the sump with carbon tetrachloride. Then replace the base plate, using a new gasket, top up with clean oil after changing the seal, and dehydrate the compressor before putting it back into use. If time does not allow for dehydration, the compressor must be purged through with refrigerant, as described on p. 52 to expel air, and a 'drier' inserted into the system. Where possible, too, it is advantageous to run the compressor for a period before opening up to the system, to give the seal a chance to settle in.

LUBRICATION

Lubrication in a small compressor is usually achieved by the eccentrics splashing into the sump oil as they rotate, so obviously the level must be adequate for this to occur. Where no visual means is provided, an assessment of the level must be made via the filler plug, using a clean dip rod, and, as a guide, the level should be about 35 mm ($1\frac{1}{2}$ in) below the main shaft. Whilst it is dangerous to run with insufficient oil, too much can also create problems, especially when the back pressure is low.

EXCESS OIL

Excess oil is often pumped through the reeds in these circumstances, which apart from the damage it may cause to them, may interrupt the proper operation of the plant.

Remedy

Surplus oil may be removed from a compressor by shutting both service valves to the system, removing the head gauge and filler plug, and drawing off excess oil by means of a syringe, testing with a clean dip rod until the correct level is reached. A compressor should be capable of going down on a 500 mm (20 in) vacuum without drawing up its oil, and as a test, put the free end of the head gauge line into an empty bottle, and run the compressor in brief spurts until the vacuum mentioned has been reached. Any excess oil will be expelled into the bottle, but should this occur in any quantity, the machine must be switched off immediately, and restarted only after the oil has drained. When a 500 mm (20 in) vacuum can be reached without oil being discharged, the compressor may be opened up to the system and put into use.

DIRECTION OF ROTATION

The direction of rotation of a compressor must be mentioned. It is not important as far as the pumping action is concerned, but there are two possible reasons why one direction would be advocated. One is in relation to the method of internal lubrication, where one direction permits a more ready flow of oil to the eccentrics, and the other concerns certain shaft seals which tend to unscrew if rotated in the wrong direction. The bellows shaft seal described is not affected in this way. Any definite direction of rotation is marked, but in general a compressor rotates in a clockwise direction, when viewed from the flywheel end.

REMOVING THE FLYWHEEL

The flywheel is often a difficult item to remove, and well-fitting pulley removers are essential for this purpose. Even these are not always immediately successful, but a sharp tap with a hammer on the end of the *removers*, whilst under tension, will do the trick. If this makes no impression, drop some liquid refrigerant slowly, a drop at a time, on to the shaft and flywheel boss, until both are well frosted. Now leave for a few minutes (the removers still on tension) and generally, without any further attention, the flywheel will be released. This is due to the temperature of the flywheel rising more rapidly than the shaft, thus unequal expansion takes place, and at the point of release the flywheel has expanded a small fraction, whilst the shaft is still unaffected.

On no account try to remove the flywheel by hammering on to the *shaft end*. This action will almost certainly injure the threads, apart from any other damage, rendering it impossible to refit the flywheel afterwards, and also remember that a flywheel is cast, and therefore brittle, and quite easily broken. Do not attempt, either, to remove a flywheel by hammering a wedge between the wheel and the seal end plate. This will bend the shaft, or cause internal damage to the connecting rods, sometimes breaking them. In fact the use of a hammer is to be discouraged, for whilst it is a most necessary and useful tool, it can be a menace if used indiscriminately.

CHOICE OF OIL

The oil in a compressor is of special quality. Apart from its lubricating qualities, it must mix readily with the refrigerant being used, and since certain quantities will circulate through the system with the gas, it must be capable of withstanding the low temperature of evaporation. Standard refrigerant oils are normally safe to use down to $-40°$, but below this, low temperature oil is required, otherwise 'waxing' or solidifying takes place in the expansion valve, blocking the orifice and preventing the flow of refrigerant. Most manufacturers specify the type of oil suitable for their units, and this should be adhered to, particularly when adding oil to hermetic units.

It is not within the scope of this book to describe the stripping down and overhaul of a compressor, nor is it something that a beginner would be expected to undertake. When a compressor knocks to the extent that workshop attention is obviously necessary, a replacement is usually fitted, the old one being renovated and put into the store for future use. The points which have been discussed, however, are all of value to the service man on site, ones which he will be frequently confronted with, and ones which every apprentice must understand.

11

The Bellows

EXPANSION

A most important feature in engineering is the expansion and contraction which takes place in substances when they are subjected to temperature changes. There are many instances where allowances have to be made for this factor, such as bends in steam pipes, gaps in railway lines, and so on, and it is assumed that every apprentice has some knowledge of this characteristic. As explained earlier, this book is intended to deal with practical rather than theoretical aspects of refrigeration, and it is sufficient for our purpose at the moment to be aware of expansion, and to appreciate that some substances expand or contract more than others, although exposed to the same temperature variations.

Expansion creates many problems in designing engines; on the other hand, it is very useful as a basis for control or information purposes, as for example in a thermometer. The heart of most controls used in refrigeration is a bellows, which is constructed of thin corrugated copper, and often used with a capillary tube attached as illustrated in Fig. 24. After being carefully charged with a gas or fluid (the type of charge depending on the use for which it is intended), the whole is sealed, and when the bulb or phial is subjected

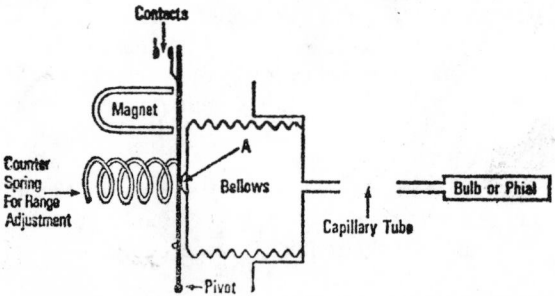

Fig. 24. Bellows and capillary. Changes of pressure within the bulb transfer through the capillary to the bellows. This in turn causes the electrical contacts to 'make' and 'break'.

to a temperature change, the charge expands or contracts. This is transferred into movement of the bellows at point A, which, in turn, is used to provide mechanical action. Charged bellows are used in all cases where change of *temperature* is the prime mover, as for example in a thermostat, but where change in *pressure* is the prime mover, then the bellows are not charged but made with provision for a pipe connection (Fig. 25). In either case, the pipe or phial allows for the actual control to be several feet from the point causing the action, and this can be seen in most coldrooms, where the thermostat phial will be found in the room whilst the instrument itself is located outside.

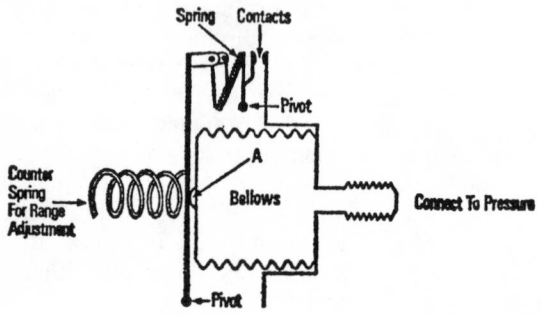

Fig. 25. 'Open' bellows for pressure control. Changes of pressure within a system are transferred to the bellows by means of a pipe to the connection shown. Note the 'snap' action provided to the contacts by the toggle and spring. The main spring, being under compression, balances the pressure in the bellows.

The following is a list of controls, together with a brief description of each, all of which employ bellows:

THERMOSTATIC EXPANSION VALVE (TEV)
Has a charged bellows, and its purpose is to meter liquid refrigerant into the evaporator, the flow being regulated by a needle valve actuated by pressure in the bellows, the pressure being regulated by the phial clamped to the suction or outlet pipe of the evaporator.

THERMOSTAT
Has a charged bellows, the action being as for the TEV except that the bellows movement operates a switch to stop and start the plant and so maintain the required temperature.

LOW PRESSURE CONTROL (LP)
This has an 'open' bellows the purpose of which is to stop the plant when the back pressure drops to a pre-set amount, and to

start it again when the pressure rises. An LP control is an extremely useful one, but it can only be used for temperature control in conjunction with a TEV, for it is the closing action of the TEV which reduces the flow of refrigerant and thereby reduces the 'back pressure'. Since pressure and temperature are related, an LP control is often used in place of a thermostat, but where a thermostat is employed as well, an LP control still fulfils a vital function in that it will stop the machine in the event of refrigerant loss. Without this it would continue to run, with the consequent risk of taking in air and moisture.

HIGH PRESSURE CONTROL

This has an open bellows and is very similar to the LP version. Its purpose is to stop the plant in the event of excessive discharge or 'head' pressure. This protection must always be provided on water-cooled units where the risk of water failure is a possibility.

WATER VALVE

This has an open bellows, connected by means of a copper tube to the compressor discharge, and its purpose is to increase the water flow as the head pressure rises. Water flow is reduced when the head pressure drops, and shut off altogether when the machine stops.

DIFFERENTIAL

The above controls are adjustable to give the pressure or temperature conditions required, and the word 'differential' is worth mentioning at this point, because of its importance in this respect. When setting an LP control, for example, it may be set to stop the machine when the suction pressure reaches 0·3 bar (4 lb), and to start it again say at 1 bar (16 lb). The difference between the two is known as the 'differential'.

It will have been noticed that three of the above controls involve the making and breaking of an electrical contact, and it is essential that a 'snap' action be obtained. Without this, as the contacts slowly parted or connected, the current would jump or 'arc', causing the contacts to weld together. Some controls use a spring and toggle action as illustrated in Fig. 25, whilst others use a magnet as shown in Fig. 24. The contact is drawn down rapidly at the last moment as the contact arm is influenced by the magnetic field, and is then held on until the pull of the bellows forces it to break suddenly.

These are the basic features of all controls, and every apprentice should read literature and study diagrams supplied by the manufacturers, and of course observe controls in actual operation whenever possible.

12
Accessories

There are a number of accessories to be found in most refrigeration systems, which whilst not essential to the effective operation of a plant, nevertheless perform a vital function. One of these is the filter drier, illustrated in Fig. 26, the purpose of which, as its name implies, is to absorb any moisture in the refrigerant, and to filter it at the same time.

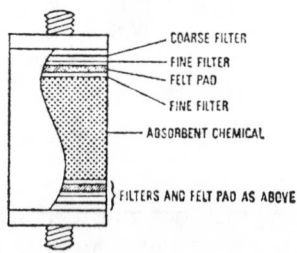

Fig. 26.   Filter drier. The chemical absorbs moisture which may be contained in the refrigerant.

DRIER

The importance of preventing traces of dirt or moisture from entering a system cannot be over-emphasised, and it is common practice to insert a filter drier at the initial installation. This should not be necessary, of course, if the engineer has used the normal procedure of using a vacuum pump, and taken care with his connections, but owing to the importance already stressed, the fitting of a drier at this stage is justified. A drier is of greatest value when moisture is known to exist in a system, or where the risk of moisture entry has been incurred when a component such as an expansion valve has been changed.

Driers are made in various sizes, but they are all similar in principle, consisting of a brass container, coarse and fine filters as indicated, felt pads, and chemical. Some driers have a set of filters at both ends, and these can be inserted for refrigerant flow in either direction. Where filters are fitted at one end only, then the drier must be inserted as indicated by the arrow on the casing. Once a drier of the twin filter type has been installed in one particular direction, however, and then for some reason temporarily removed, it must be replaced in the same direction as before, otherwise any impurities trapped in the entry filter are liable to be discharged back into the system.

The chemical is usually either activated alumina, silica gel, or molecular sieves, each capable of absorbing moisture which may be contained in the refrigerant, and known as 'desiccants'. Molecular sieves have a higher moisture-absorbing capacity than the other two, and are commonly used in modern plants using R.12 or R.22. They are also more stable. Activated alumina, for example, will on occasions break down, perhaps through vibration, allowing a fine white powder to enter the system with disastrous results. For this reason a drier must always be securely fixed and also placed in a perpendicular position with the liquid made to flow downwards. Molecular sieves must not be used with methyl chloride refrigerant.

SIGHT GLASS

Another important item is the sight glass (*see* Fig. 27). This is made of brass and has a small glass 'window' which permits the flow of refrigerant to be observed. When the system is fully charged

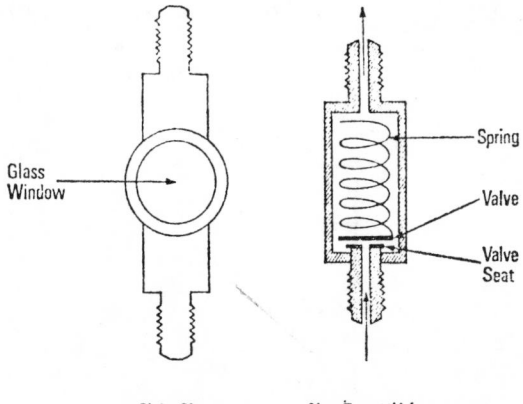

Sight Glass   Non-Return Valve

Fig. 27.   A sight glass enables the flow of refrigerant to be observed, and gives visual indication of gas shortage. The non-return valve permits refrigerant flow in one direction only.

and the plant running, the sight glass is 'clear', but when short of refrigerant, bubbles will be seen. The rate of bubbling indicates to some extent the degree of refrigerant shortage. The sight glass, for convenience sake, is usually fitted after the drier near the unit, but where liquid lines are long, say 15 m (50 ft) or more, the sight glass is better located near the expansion valve, otherwise it sometimes gives a false impression of gas shortage when the system is in fact fully charged.

FILTER

A filter, consisting of very fine brass gauze, contained in a brass container, is another useful addition in a system. It may be inserted in the liquid line to prevent foreign matter from the unit reaching the expansion valve, or in the suction line to prevent particles from the evaporator entering the compressor. The latter is of particular importance in new installations where scale from brazed joints may still be in the suction piping.

HANDWHEEL VALVES

These are shut-off valves inserted into a system and operated by hand. The flow of refrigerant may now be isolated or diverted as required. They are used in a variety of ways, particularly in installations of 2 kW (3 hp) and above. Figure 28 shows one such purpose which enables the drier to be isolated and removed with the minimum of difficulty. They may also be used immediately after the liquid receiver for charging purposes. Whereas in small plants it is usual to charge refrigerant through the compressor suction as a

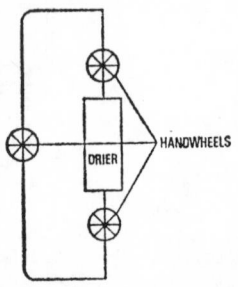

Fig. 28. Handwheel or shut-off valves may be used for many purposes. The illustration shows how a drier may be isolated and by-passed.

gas, by the use of a 'charging' valve, liquid refrigerant can be allowed to flow in, thereby saving time. When the weight of refrigerant is perhaps 20 kg (50 lb) or more, this method of charging is most necessary.

NON-RETURN VALVE

This is a valve which can be inserted into a system, which permits the flow of refrigerant in one direction only (*see* Fig. 27). These are particularly necessary in the 'reverse-cycle' defrosting system as described on p. 110.

HEAT EXCHANGER

Mention must also be made of the heat exchanger, which is commonly fitted in systems of about 0·75 kW (1 hp) and upwards. This consists of a brass container, made in such a way that the warm liquid refrigerant is in contact with the cool suction pipe from the evaporator. The heat exchanger therefore achieves the dual purpose of pre-cooling the liquid before evaporation, and of reducing the possibility of 'frosting back' on the suction piping. Figure 29 illustrates the principle, and it is advisable to study literature issued by manufacturers, which will provide greater detail.

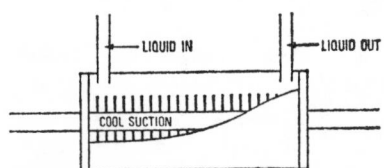

Fig. 29. Heat exchanger. Cool gas from the evaporator is used to reduce the temperature of warm liquid refrigerant before it enters the expansion valve.

As already mentioned, each one of these components is additional to those required to perform refrigeration, but on the other hand they play an important part in ensuring trouble-free operation. It is suggested that every student should make a point of observing such items in use, and of querying any with which he is not familiar.

PROGRESS CHECK

This is a convenient stage to remind every apprentice of the practical features mentioned earlier in the book, which are so important in refrigeration service. Remember to plug gauge lines and gas cylinders immediately after use, and do not leave copper tubing exposed to the atmosphere. This applies to the system itself, as for example when changing a drier or expansion valve, or to new copper tubing being used during an installation. No apology is made for repeating these elementary instructions, for they are so often neglected, even by experienced men.

Also be well equipped with tools, and learn to look after them. Take special care of gauges, protect them from becoming damaged, and check them regularly for accuracy. Always be in possession of a thermometer, and if it gets broken, renew it immediately, for it is impossible to service refrigeration without one.

TEST QUESTIONS

On p. 141 Test Questions No. 2 are to be found. These cover the matters dealt with since Test Questions No. 1, and the student is advised to be fully conversant with the answers before proceeding.

13
Looking and Listening

COMMON PROBLEMS

Before proceeding to the electrical side of refrigeration, which of course is most important, it is intended as an exercise to pose a few problems common in refrigeration, based on the facts dealt with so far. In this way the student can check on his own ability to diagnose a fault, and also learn something of the method of going about it. He should not find this unduly difficult, if, in addition to studying other literature, he has asked plenty of questions and made observations of his own.

COLDROOM TEMPERATURE—HIGH

For this purpose, suppose that a butcher has complained that the temperature of his coldroom has gone up, and you, the apprentice, being the only person available, have arrived on the scene complete with tools and spares to deal with the situation. It is a 0·75 kW (1 hp) open type condensing unit, operating a forced draught evaporator. The unit is running. Well, where do you go from here?

TESTING
The very first thing is to avoid hasty conclusions, especially on the evidence of the butcher himself. He is a busy man, and whilst any information he may impart is extremely valuable, it is not always accurate. Therefore listen by all means, but use such enlightenment in conjunction with your own knowledge. Careful inspection is the key to all accurate diagnosis, establishing facts, and narrowing the possible causes down to a few, checking all the while. Look for the obvious and easy solutions rather than fearing the worst. Do not immediately rush to put on gauges, for example, but first place a thermometer into the coldroom and then have a slow methodical

look-round, absorbing every detail. Note whether the belt is slack, feel the suction and discharge pipes from the compressor and the liquid line from the receiver. When condensing correctly, the suction should be cool, the discharge hot. The liquid line should be warm, and some warmth is normally felt beyond the drier, but since there is a fault on this particular unit, a clue to the cause might be found in these tests. Also notice if the condenser is clear of dirt and fluff, and check if the air flow temperature through it appears to be high. Look for any traces of oil leaks, and feel both sides of the drier or filter if fitted. A partial restriction in either of these may create a pressure drop sufficient to cause cooling or even frosting of the outlet pipe. See if the condition of the electric leads is satisfactory, and make a mental note of the controls being used. Having inspected a unit in this way, go into the coldroom, check the temperature, see if the fan is running, look at the evaporator to see if the coils are clear of frost, or if excessively frosted, and listen for any hissing at the expansion valve.

This preliminary examination need not take more than a few minutes, but to the trained engineer provides vital information which he will automatically summarise into facts. Naturally, if the belt is very slack, the compressor stationary, the plant or evaporator fan not running, or the condenser choked, the fault is fairly obvious; but let us consider a few lines of thought based purely on 'looking and listening', which might apply to the case we are discussing.

FACTS

Compressor running. No sign of belt slip. Condenser clear. Unit site reasonably cool. Coldroom fan running. No frosting taking place on the evaporator. No sound from the expansion valve. Compressor discharge giving no feel of heat. Compressor suction giving no feel of cooling. Liquid line at ambient temperature.

This condition would indicate that there was no flow of refrigerant through the system.

POSSIBLE FAULT

System out of refrigerant. System blocked. Compressor reeds broken.

Since broken reeds are not very usual, one of the other two is the most likely.

If the system is blocked at some point, no refrigerant will be circulating, and it follows that the suction side of the compressor will, in this case, be on a vacuum. Therefore, after checking that the suction service valve is 'back seated', and fitting the gauge line, do not open up to the system without first 'purging' the gauge line with a puff of gas from the refrigerant cylinder, and then ensuring that the connections to the compressor and to the gauge are tight. Without this precaution, air in the gauge line may be drawn into the system, thus causing further complications.

BLOCKAGE

Having fitted both gauges carefully and 'cracked' both service valves just enough to give a reading, all the further information required will now be available. A blocked system will be indicated by the suction gauge registering a vacuum of 500 mm (20 in) or more, and confirmation of this will be obtained by a pressure being shown on the head gauge. In this case stop the plant and if the suction pressure does not rise, put refrigerant gas from the service cylinder into the

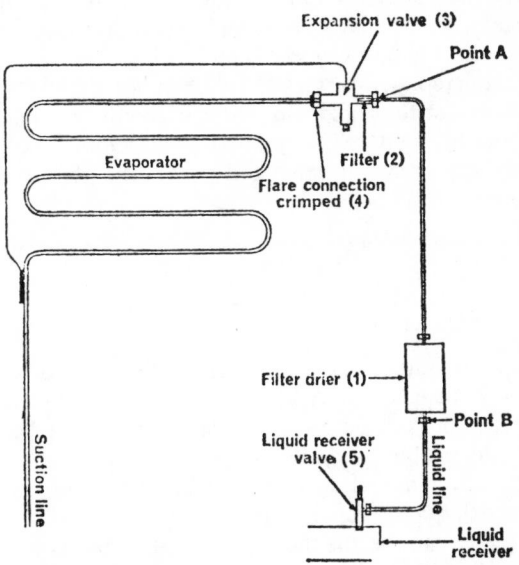

Fig. 30. Points of blockage in a refrigeration system. The illustration indicates the points at which a blockage is most likely to occur.

suction to bring the pressure to atmospheric. By reference to Fig. 30, a gas pressure of atmospheric will now exist in the system up to the blockage, with a pressure either of gas or liquid on the other side. The most likely places to cause a choke are as follows:

1. Drier or filter.
2. Expansion valve filter.
3. Expansion valve orifice.

4. A 'crimp' or crushing of a copper tube in a flare-nut (most likely on the outlet of the expansion valve).
5. Liquid receiver valve.

Now it is only necessary to slacken each flare-nut from the expansion valve and check to see if a pressure of gas or liquid is released. For example, if liquid is shown to be present at the inlet flare connection to the expansion valve as marked at point A, it would obviously indicate a choke at either 2, 3 or 4. If no gas pressure exists at this point, then tighten the flare-nut again, and go back to point B which is the unit side of the filter-drier. On most occasions, any choke which exists in a system of this kind will be located in one of these two items. Having established which one, shut the liquid receiver valve, release the liquid pressure to the atmosphere, and change the faulty part or reflare the copper tube, as the case may be. It will not be possible to save the liquid refrigerant trapped between the blockage and the liquid receiver valve, so after carrying out the repair, insert a sight glass (if not already fitted), open up the system, test for leaks and add refrigerant to replace the lost charge. When this is complete, 'pump down', remove the sight glass, open up the liquid receiver valve and restore the plant to its correct operation.

LEAK

If the system is out of refrigerant, then little or no reading will be shown on the gauges. A leak of this magnitude indicates a fractured pipe, broken flare-nut, or burst shaft seal bellows, and it is readily detectable by allowing gas into the system, and either listening or using a leak detector. After attending to the leak, insert a new drier, and recharge using a sight glass as before.

Broken reeds can be diagnosed by testing the compressor for pumping as described in Chapter 11, and attending to it accordingly.

Let us have another inspection based on 'looking and listening' as before.

FACTS

Compressor running. No sign of belt slip. Condenser clear. Unit site cool. Coldroom fan running. Slight frosting taking place on the evaporator. Expansion valve hissing. Compressor discharge warm but not hot. Liquid and suction pipes giving no tangible indication of heat or cooling. Trace of oil on drier flare connection.

POSSIBLE FAULT

Shortage of refrigerant. (Trace of oil a clue to the leak.)

The fault of gas shortage is readily confirmed by the head gauge pressure showing lower than normal, 'normal' being ascertained from the pressure/temperature charts. Further confirmation can be obtained from the sight glass, and if a sight glass is not fitted, then 'pump down' and insert one. Now test for leaks with the test lamp, using any traces of oil as clues, and after locating and repairing the leak, recharge the system, again using the sight glass.

One final check.

FACTS

Compressor running. No sign of belt slip. Condenser clear. Unit site cool. Compressor discharge hot. Liquid line warm. Compressor suction frosted. Coldroom fan running. Evaporator excessively frosted.

OVER-FROSTING

This condition indicates loss of control by the thermostatic expansion valve.

POSSIBLE FAULT

Evaporator badly in need of defrosting. TEV phial not correctly clamped to the evaporator. TEV out of adjustment or stuck open. Compressor losing efficiency.

The remedy to the first two faults is obvious, and with regard to the third, if there is any doubt about the TEV it is usually quicker and more satisfactory to replace it. Of course the system must be 'pumped down' before this can be done, and when removing the valve take particular care to avoid moisture entering the pipes by carefully drying the area round it beforehand. Even with this precaution it is common practice to replace or fit a new filter drier into the circuit at the same time.

Referring to the fault of 'compressor losing efficiency', very occasionally a plant will produce refrigeration only when the 'back pressure' is high, and if in doubt regarding the compressor's efficiency, test for pumping as described.

The main purpose of this chapter has been to show how much information can be obtained by merely 'looking and listening', without any tools or instruments beyond a thermometer, and how important it is to muster all the facts at the very beginning. Not only that, but to record them on the work sheet so that comparison can be made after adjustments have been carried out.

14

Elementary Electrical Notes

It is now necessary to consider electricity, which is employed almost universally to power mechanical refrigeration. There are several main reasons for this. First, it is clean and does not produce fumes, the power units are compact and quiet in operation, but its chief virtue probably is its adaptability to automatic control. Whether it is a domestic or a large coldroom, the load on a plant varies with its use and ambient temperature, and naturally it is beneficial for the machine to be stopped and started as the load demands.

Most domestic refrigerator breakdowns are due to electrical faults, and in fact much of the servicing required on them can be carried out only by men with electrical knowledge. On larger installations the wiring will normally have been executed by specialist electricians, but when the plant fails, no matter what the reason, it is the refrigeration engineer who will be called. Often he is unfamiliar with electrical problems, and to avoid being embarrassed in this way, a beginner must equip himself with a sound working knowledge of electricity. It is an integral part of the refrigeration trade, and where the local technical college cannot offer a course in refrigeration, a course in electricity might well be taken instead.

There are many applications using single phase motors up to, say, 0·75 kW (1 hp) and at least half the faults that occur on them can be dealt with by the engineer on site. Therefore a knowledge of small electric motors is also essential and, as stated before, this book will deal with the practical aspect of such matters, rather than the theoretical. It is advised that the student will study theory in conjunction with them.

UNITS OF MEASUREMENT

However, it is necessary to mention the familiar measures of electricity which are volts, amperes, ohms and watts. These are easily understood if electricity through a circuit is likened to water passing through a pipe. Volts are then represented by the pressure of water, amps by the quantity or flow of water, ohms the resistance in the pipe to the flow, and watts the amount of water being used. The mains voltage or pressure remains constant, so in order to allow a greater flow of current, a larger size of wire would be necessary, just as to increase the flow of water a larger bore pipe would be needed.

WIRE AND FLEX

The type of metal wire and its length also affects resistance, some being better conductors than others. Copper is extensively used in cables because of its good conductivity, in addition to which it is non-ferrous, non-magnetic, ductile, and readily soldered. If the wire is too small to carry the current passing through it, it will get hot, and this feature is exploited in 'fuses' to protect the circuit. Fuses will melt and break the circuit if subjected to a current in excess of their designed capacity.

Electrical measures are related in the following way:

If V = Volts, I = Amps, R = Ohms and W = Watts

Then

$$I = \frac{V}{R} \qquad V = R \times I \qquad R = \frac{V}{I}$$

$$W = V \times I \qquad V = \frac{W}{I} \qquad I = \frac{W}{V}$$

Thus if any two are known, the remainder may be found. In practice the current consumption of an apparatus is marked on it. The capacity of cables and flexes is given in manufacturers' tables; therefore it is a simple matter to select a cable adequate in current carrying capacity for the job in hand. As an illustration, let us assume an electric fire is rated at 3 000 watts. What current will this consume?

From the above we have

$$I = \frac{W}{V} = \frac{3\ 000}{230} = 13 \text{ amps (approx.)}$$

Therefore a cable of 13 amps carrying capacity would be required.

It is interesting to note that as the normal power point is rated at 13 amps, this is the maximum load which can be placed on the same plug top.

VOLTAGE

Standard electricity supply throughout the country is 220/240 volts, alternating current (AC) single phase. 'Single phase' briefly means that only one of the three phases being produced by the generator, and dispatched by the power station, is used. For more powerful plant, such as motors of 0·75 kW (1 hp) and above, all three phases are used, the voltage then being 400–440.

The wiring for standard single phase is carried out in three-core cable, brown for 'live', blue for 'neutral', and green and yellow for 'earth'. Sometimes the 'earth' is a separate bare copper wire, and since its purpose is for safety only, it is connected directly to the metal casing of the appliance. Figure 31 illustrates a test lead which

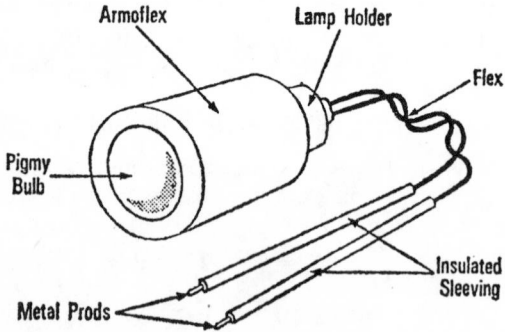

Fig. 31. Test lead. With this simple instrument it is possible to check that current is reaching the plug, and to ensure the plug is correctly wired. A test lead *must* be fused and adequately insulated. Fuses are in the insulated sleeves. Prods suitable for this purpose are obtainable from electrical suppliers, and must be used in conjunction with the insulated lead and lamp holder illustrated.

every apprentice must possess, consisting of an ordinary bakelite lamp holder, a pigmy bulb, protected by a few inches of Armoflex, and rubber-covered cable soldered into two prods as shown. The prods are specially made for the purpose, and apart from being adequately insulated, contain, within the body, replaceable cartridge fuses.

With this instrument it is possible not only to check that current is reaching the socket, but also to ensure that the socket is correctly

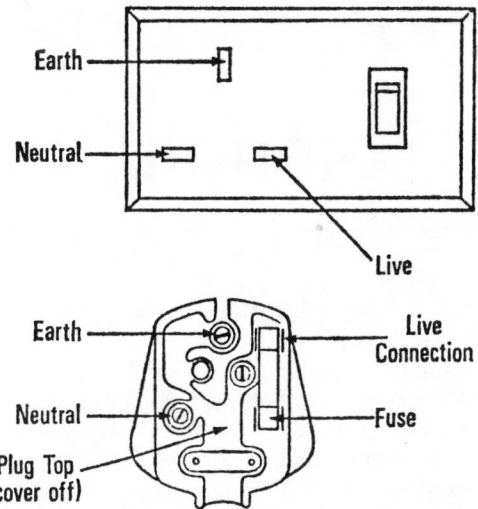

Fig. 32. Plug top and socket. Always ensure that the 'live' connection is correctly wired.

wired. 'Live' should be on the right-hand connection as shown in Fig. 32, and a light will be obtainable not only between the brown and blue leads, but between the brown and earth leads. This establishes the 'live' lead. No light is obtainable between neutral and earth.

The plug top must be wired to provide a live connection to the corresponding right-hand pin, for although the apparatus will operate equally well whichever way the leads are placed, switches and controls are always wired to cut off the 'live' connection. This may appear to be elementary information to many, but a plug top is so often incorrectly wired that it is obviously necessary to state these simple facts.

15
The Solenoid

THE MAGNETIC FIELD

When a current of electricity is passed through a coil of wire a magnetic field is created, which ceases immediately the current is switched off. This feature is exploited extensively throughout electrical control circuits, and the electric bell, illustrated in Fig. 33, is a simple example. When current magnetises the iron core, arm A is drawn towards it, but as soon as this action takes place the

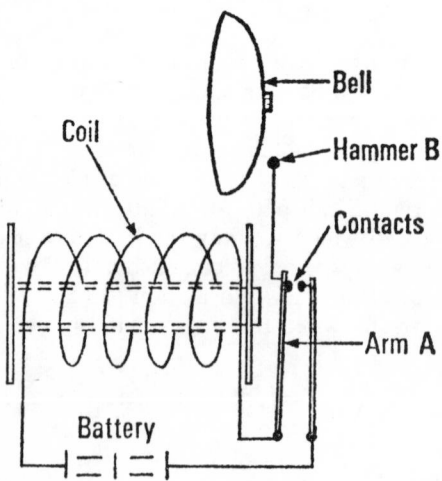

Fig. 33. Electric bell. This illustrates one use of a solenoid.

circuit is broken. Arm A therefore springs back, the movement then being repeated, thus hammer B will oscillate rapidly, producing sound from the bell.

Coils used to produce magnetic fields are known as solenoids, and valves operated by solenoids are familiar items in many refrigeration systems. Figure 34 illustrates the general working of such a valve, and shows how the principle mentioned above is utilised to achieve

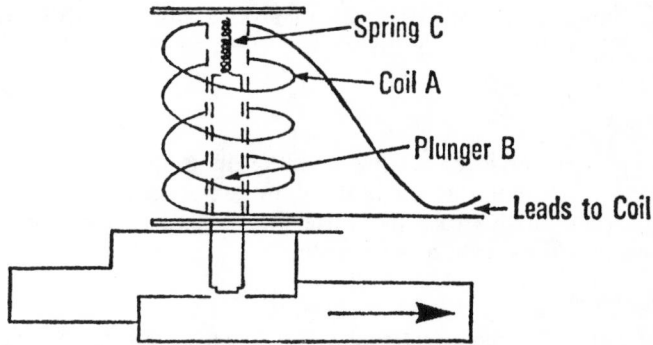

Fig. 34. Solenoid valve. A solenoid being used to operate a valve electrically. C is a compression spring.

a different purpose. When coil A is energised the magnetic field lifts plunger B from its seat, allowing the liquid or gas to flow, but when the current is switched off, the weight of the plunger, with assistance from the spring C, shuts the valve again. It can be seen, therefore, that a solenoid valve is normally shut, and the coil has to be energised to open it.

Solenoid valves are especially designed for the various purposes for which they are to be used, and where diaphragms are incorporated, these have to be of a suitable material, depending on whether for refrigerant, oil or water. Before fitting a solenoid valve, therefore, ensure it is appropriate for the duty required, particularly regarding temperature range and coil voltage, making sure the valve is fitted into the system upright with the direction of flow as indicated by the arrow. The valve will not function if fitted incorrectly. Manufacturers provide detailed drawings and descriptions of their products, which the apprentice is advised to study carefully. He should also obtain an old valve to strip down and examine for himself.

SERVICING

A useful feature worth remembering about a solenoid valve is that, unlike most refrigeration components, it is relatively easy to service. The coil, for example, is quite readily changed on site, without any prior preparations other than to switch the current off. In addition to this, by pumping the system down, or in the case of a water valve, by turning off the water supply, some makes of valve can be dismantled, and their interior parts replaced. If the seat is badly worn, however, or if other considerations make it seem preferable, then the valve should be replaced.

A common position for a solenoid valve is immediately before the expansion valve, it being situated there to perform one of the following functions.

FUNCTION 1

To prevent liquid refrigerant flooding the evaporator during the 'off' cycle, which in turn might flood the compressor when the plant is restarted. In this case the valve is energised and de-energised in unison with the compressor motor.

FUNCTION 2

To shut off liquid refrigerant to a cabinet or coldroom which is down to its desired temperature, whilst other compartments connected to the same condensing unit still require refrigeration. In this case the solenoid is operated by a thermostat controlling each cabinet independently.

FUNCTION 3

To prevent excessive pressure build-up in the evaporator, during the operation of electric defrosting. To achieve this, a time-switch de-energises the solenoid valve at the same time as current is connected to the evaporator heaters. The compressor continues running, however, but since the flow of liquid has now been stopped, the back pressure will be reduced to a point where a low-pressure cut-out stops it. Should the pressure in the evaporator rise during defrosting, the pressure control will restart the plant, thus reducing the pressure again.

Solenoid valves are always necessary with hot-gas defrosting systems. A time-switch connects current to a solenoid valve fitted into the compressor discharge pipe, which allows hot gas to be by-passed from the condenser and led direct to the evaporator. The compressor of course continues running (*see* p. 107).

Where automatic water-defrosting is employed a similar action takes place, only this time a time-switch operates a water solenoid valve, fitted into the water supply pipe, and the plant is simultaneously stopped.

STARTER

Whilst discussing the uses of solenoids in refrigeration controls, mention must be made of the solenoid coil in a three-phase starter box. A three-phase motor requires that three separate contacts be made at the same moment, and to achieve this a solenoid is employed. A starter can look a very complicated piece of apparatus, as indeed

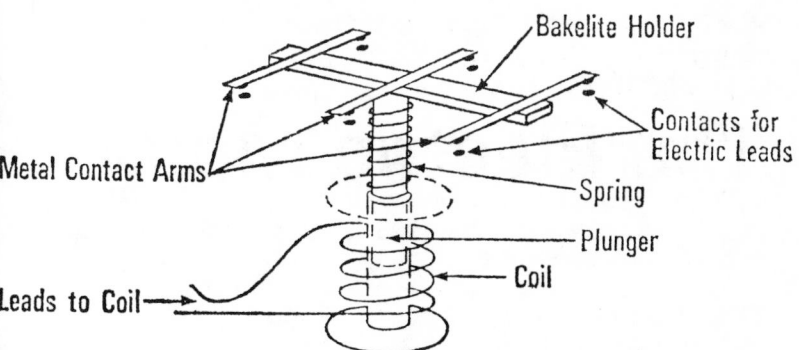

Fig. 35.   Three-phase starter. A solenoid being employed to operate three
contacts simultaneously in a three-phase starter.

it is when incorporating safety devices, reset buttons, and so on, but
basically it is just a switch, and Fig. 35 illustrates again how an
energised coil lifts a central core, enabling three separate contacts
to be made together.

RELAY

Finally, the relay, universally used to start hermetic units, also has
a solenoid, and this is described in detail on p. 91 in conjunction
with the domestic refrigerator. It should be obvious from what has
been said what an important role the solenoid plays in automatic
controls.

16
The Electric Motor

In the last chapter it has been shown that an electric current passing through a coil of wire will create a magnetic field. If, therefore, three coils are placed in a circle, and each coil is separately connected to one phase of a three-phase power supply, then as each coil reaches its maximum current successively, so the magnetic flux passes from one coil to the next. In this way a rotating magnetic field is established.

In addition to this, relating to the electric motor, the two following basic principles apply:

a. If a rotating magnetic field cuts a loop of wire, an electromotive force is induced into the conductors and a current flows round the loop.
b. If a loop of wire is placed into a magnetic field, it tends to move away from the field when a current is passed through it.

ROTOR AND STATOR

The two operative parts of an induction motor are the stator, which carries the coils, and the rotor which revolves inside the stator. Figure 36 illustrates the end view of such a stator showing the position of the coils and the rotor. The rotor is made up of iron laminations, with holes at regular intervals round the periphery as shown, through which copper or aluminium conductors are fitted, the ends of which are welded together to form a cage. Figure 37 shows the side view of a rotor.

The rotating magnetic field cuts the conductors, and induces current into them (principle a), whereupon the conductors try to move out of the field (principle b). Since the field is rotating, the conductors also rotate, the rotor never rotating as fast as the field. If it did the conductors would not be cut by the magnetic flux, and no current therefore induced into them. This difference of speeds is known as the 'slip', and varies with the load placed on the motor.

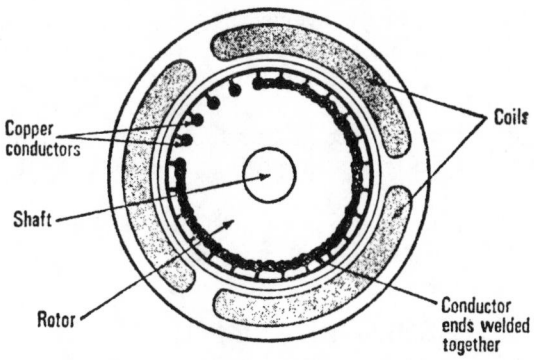

Fig. 36. Electric motor. The illustration shows the end view of the stator windings and rotor. As each field is energised consecutively, so a rotating magnetic field is created.

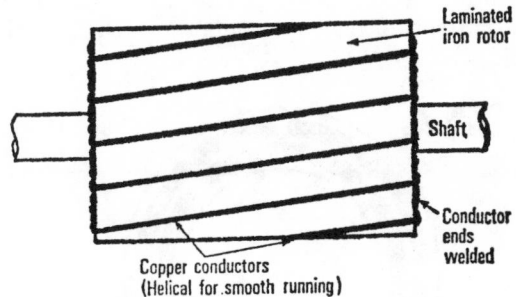

Fig. 37.   Electric motor rotor (side view). Note the conductors passing from one side to the other.

STARTING A SINGLE-PHASE MOTOR

Most motors up to 1 kW (1½ hp), and all domestic refrigerator motors, are designed to run on single-phase supply, and from what has been said, it should be clear that a corresponding single-phase motor would consist of a single coil only. This would pulsate from zero to maximum, creating a magnetic field, but it would not be rotating. The rotor conductors not being cut, therefore, they would not be energised. If, however, the rotor is given a movement by hand, the magnetic field *would* be cut, and current induced into the conductors as for three-phase. Once started, rotation continues owing to the rotor current itself setting up a magnetic field at 90° in time and space with the main field.

Obviously this is not a convenient way to start a motor, so automatic starting is devised by introducing a second winding between
the main windings, which attains its maximum at a later instant
than the main field. Once the rotor is started the secondary winding
is taken out of circuit (since it is no longer required until the next
start) by means of a relay or centrifugal switch, both of which are
described later on.

The following three single-phase motors, widely used in refrigeration, all employ the principles mentioned.

SPLIT-PHASE MOTOR

Figure 38 illustrates the windings of a four-pole split-phase motor,
and shows the 'run' windings A, and 'start' windings B. The running
windings possess high inductance and low resistance, whilst the
starting windings have relatively high resistance and low inductance.
When the current is switched on, a difference of phase is set up
between the currents of the two windings, which approximates to a
rotating field.

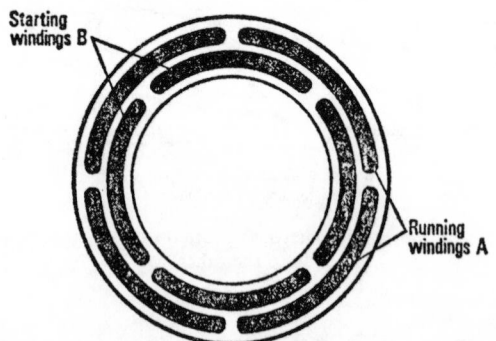

Fig. 38.   Split-phase four-pole motor, showing position of windings. Starting
windings taken out of circuit once the motor is running.

This type of motor has a low starting torque, but is very suited
to the domestic refrigerator, where the starting load is reduced
during the 'off' cycle, by the gas pressures equalising through the
capillary tube (*see* Chapter 17 dealing with the domestic machine).

CAPACITOR-START MOTOR

This motor is wound as for split-phase, except that the starting
and running windings are similar, and the difference in phase is

obtained by inserting a capacitor in series with one of them. The capacitor-start motor has a starting torque approximately twice that of the split-phase motor, and is universally used on systems employing an expansion valve.

SHADED POLE MOTOR

This has a very low starting torque, but it is extensively used for small fans. Its operation is best explained by referring to Fig. 39. The rotor is the 'cage' type as for the other two motors, but the

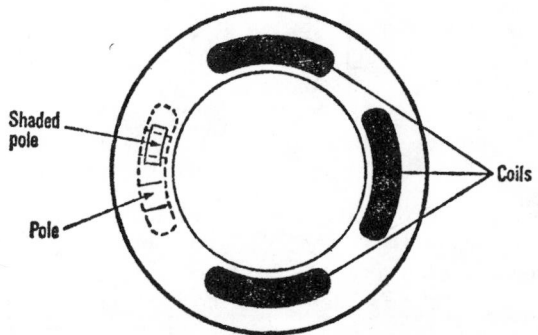

Fig. 39.   Shaded pole motor. Each winding contains two poles, one of which is 'shaded' by a copper band. This retards the growth of the magnetic field in that pole, creating a resemblance to a rotating magnetic field.

stator is wound on four poles. Each pole is split as shown, with one half 'shaded' with a copper band. The result of this is to retard the growth of the magnetic field in the shaded portion; thus there is a tendency for the field to move from the unshaded to the shaded pole, and a rotating magnetic field is created.

The student is urged to obtain some old motors of the types mentioned, and to strip them down and check for himself the facts mentioned, and in particular to study the various kinds of centrifugal switches to be found in the capacitor-start motor, the principle of which is as follows:

CENTRIFUGAL SWITCH

Figure 40 shows a centrifugal switch, which is fixed to, and therefore rotates with, the rotor. In this figure the motor is at rest and the operative part A is pressing switch contacts B so that contact is being

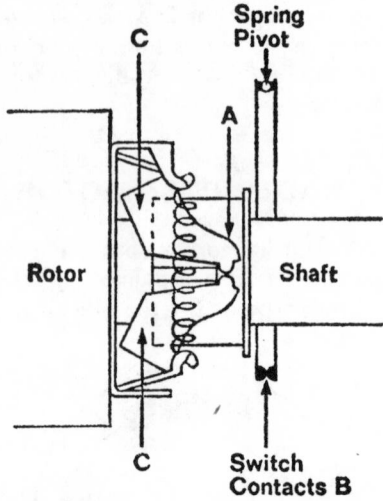

Fig. 40.   Position of switch with the motor stopped. Contacts closed.

made. This switch is in circuit with the starting windings (*see* Fig. 40). When current is switched on, the rotor revolves, and as it gathers speed, the weights C fly out by centrifugal force. This draws part A away from the switch to the position as shown in Fig. 41, and in turn the contacts, being relieved of pressure and assisted by a spring, part and so break the circuit.

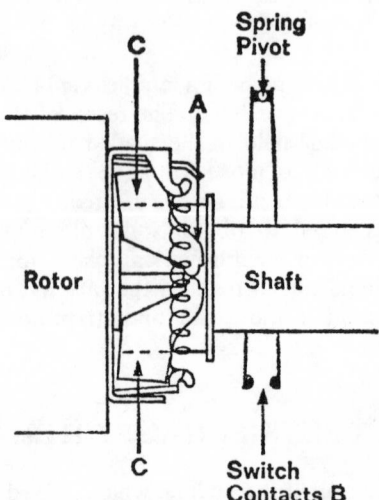

Fig. 41.   Centrifugal switch. Position of switch with the motor running. Contacts open.

TERMINAL BLOCK

Figure 42 shows a typical terminal block with wiring. The connections
may vary with different motors, the mains connections usually being
marked, but if the markings have become illegible, and the leads
removed, it is fairly easy to determine the correct terminals by
remembering that the run winding is at all times connected to the
mains, whilst the start winding has only one lead direct to the mains,
with the other passing through the switch.

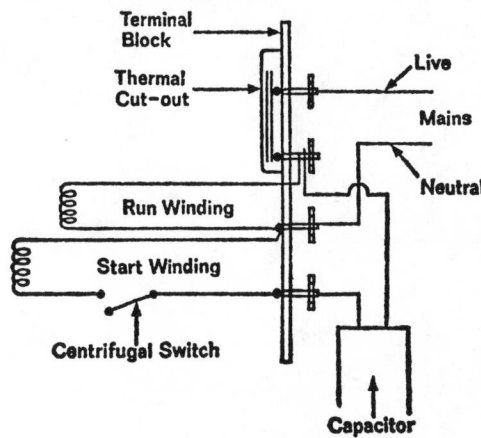

Fig. 42.   Terminal block. Wiring of a capacitor-start single-phase motor.

THERMAL OVERLOAD

The overload, indicated in Fig. 42, is illustrated in greater detail in
Fig. 43. It is fitted into the 'live' side of the mains lead, and consists
of two dissimilar metals joined together as shown, known as a
bi-metallic strip. The heater wire surrounding the strip, if subjected
to a current in excess of normal for the motor, will heat sufficiently
for the metals to expand. Being of different metals, expansion is
unequal, the strip bends, and in so doing the contacts are parted.

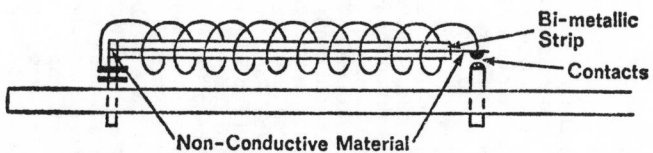

Fig. 43.   Thermal overload. Two dissimilar metals joined together expand
unequally when heated, causing the arm to bend and break contact.

Fig. 44. Single-phase capacitor-start motor. This drawing clearly identifies the component parts of a fractional horse-power motor. The centrifugal switch, it should be noted, differs from that described in Figs. 40 and 41. Each manufacturer's motors vary in detail but the principles remain the same.

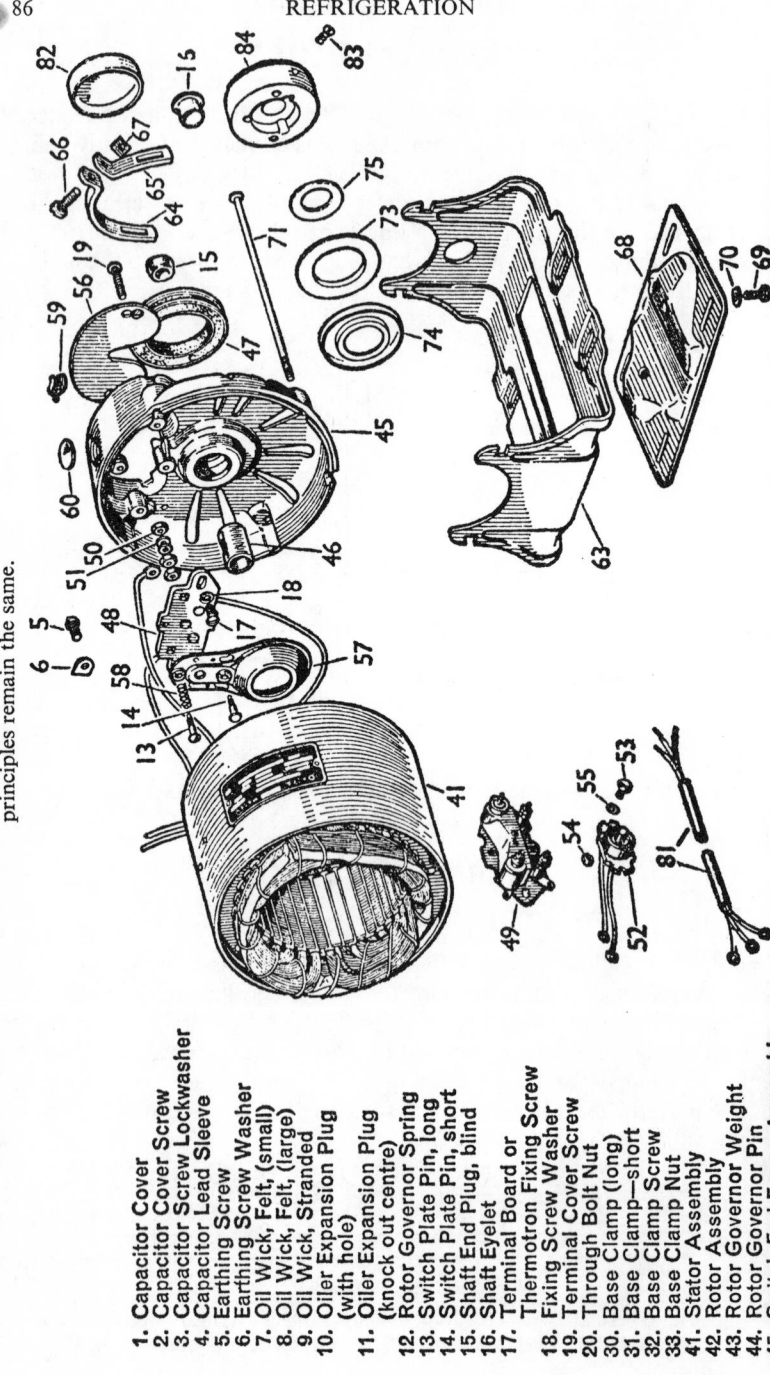

1. Capacitor Cover
2. Capacitor Cover Screw
3. Capacitor Screw Lockwasher
4. Capacitor Lead Sleeve
5. Earthing Screw
6. Earthing Screw Washer
7. Oil Wick, Felt, (small)
8. Oil Wick, Felt, (large)
9. Oil Wick, Stranded
10. Oiler Expansion Plug (with hole)
11. Oiler Expansion Plug (knock out centre)
12. Rotor Governor Spring
13. Switch Plate Pin, long
14. Switch Plate Pin, short
15. Shaft End Plug, blind
16. Shaft Eyelet
17. Terminal Board or Thermotron Fixing Screw
18. Fixing Screw Washer
19. Terminal Cover Screw
20. Through Bolt Nut
30. Base Clamp (long)
31. Base Clamp—short
32. Base Clamp Screw
33. Base Clamp Nut
41. Stator Assembly
42. Rotor Assembly
43. Rotor Governor Weight
44. Rotor Governor Pin
45. Switch End Frame Assembly

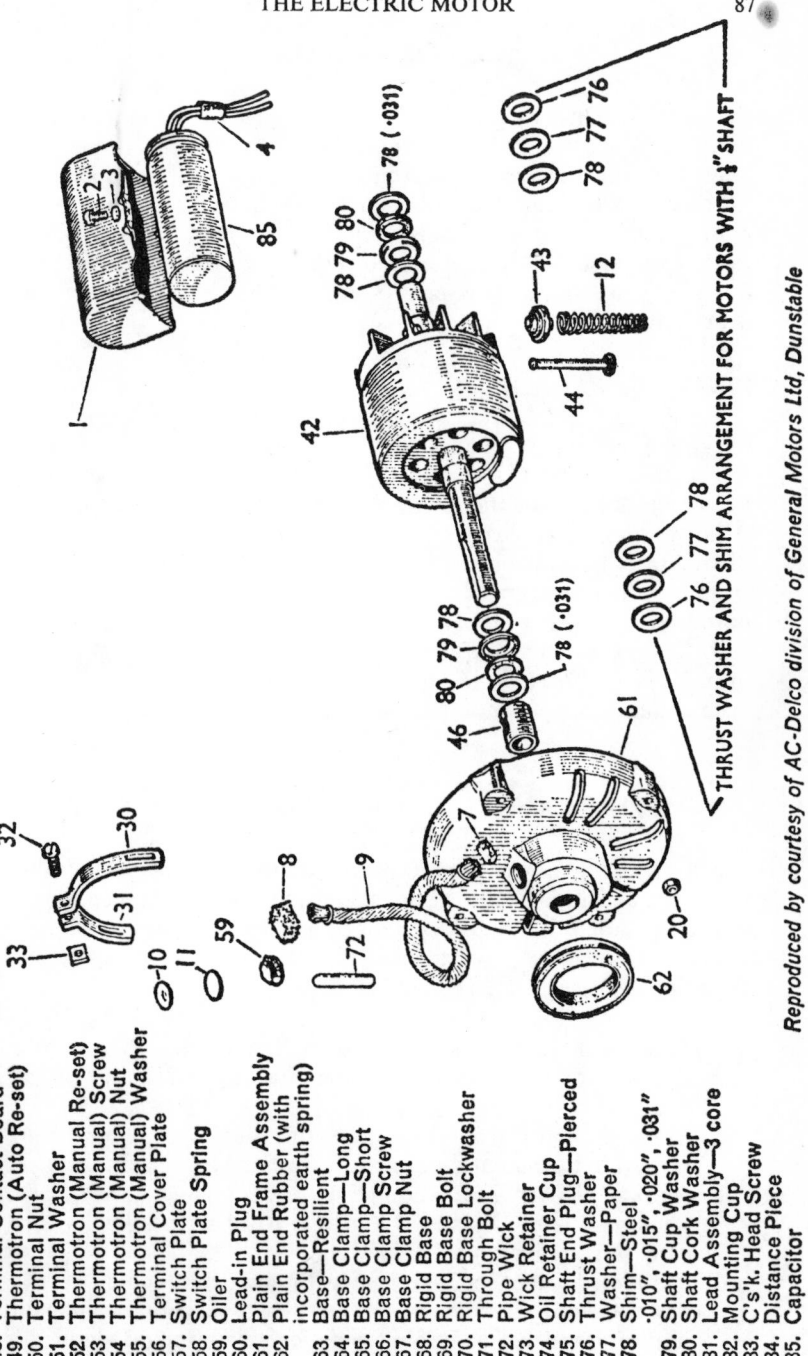

THRUST WASHER AND SHIM ARRANGEMENT FOR MOTORS WITH ⅜" SHAFT

Reproduced by courtesy of AC-Delco division of General Motors Ltd, Dunstable

48. Terminal Contact Board
49. Thermotron (Auto Re-set)
50. Terminal Nut
51. Terminal Washer
52. Thermotron (Manual Re-set)
53. Thermotron (Manual) Screw
54. Thermotron (Manual) Nut
55. Thermotron (Manual) Washer
56. Terminal Cover Plate
57. Switch Plate
58. Switch Plate Spring
59. Oiler
60. Lead-in Plug
61. Plain End Frame Assembly
62. Plain End Rubber (with
    incorporated earth spring)
63. Base—Resilient
64. Base Clamp—Long
65. Base Clamp—Short
66. Base Clamp Screw
67. Base Clamp Nut
68. Rigid Base
69. Rigid Base Bolt
70. Rigid Base Lockwasher
71. Through Bolt
72. Pipe Wick
73. Wick Retainer
74. Oil Retainer Cup
75. Shaft End Plug—Pierced
76. Thrust Washer
77. Washer—Paper
78. Shim—Steel
    ·010" ·015" ·020", ·031"
79. Shaft Cup Washer
80. Shaft Cork Washer
81. Lead Assembly—3 core
82. Mounting Cup
83. C's'k. Head Screw
84. Distance Piece
85. Capacitor

MOTOR FAILURE

Evidence of motor failure is often indicated by a periodic 'hum', followed by a 'click' as the overload curtails the current. In the type of motor being discussed some external factor may well be responsible for this, such as a seized compressor, fouled pulley, or of course the motor bearings themselves being worn or seized. These are possibilities to examine first and deal with accordingly. But where the motor is free to rotate and the above conditions exist, then the starting circuit is at fault.

PROBABLE FAULTS

A quick and simple test is worth trying. Switch the motor current off, rotate the pulley by hand a few times, and then switch on again. Repeat this action, and if the motor can be made to start the centrifugal switch is at fault, the movement having re-established contact. If no response is received pull the mains plug out, disconnect the capacitor leads, and temporarily connect in a new capacitor. Should the motor now start, the capacitor is obviously the trouble, and the new one can be fitted in permanently, but if not, then again a fault with the internal centrifugal switch is indicated. The point to bear in mind is not to strip the motor down until all possibilities have been eliminated.

TO STRIP DOWN A FRACTIONAL HP MOTOR

When stripping down a motor the following points should be observed, and Fig. 44 will assist in identifying the parts mentioned. (It will be noted that this particular make of motor uses a different pattern of centrifugal switch from that illustrated in Fig. 40, but the principle is the same.)

1. First ensure that the power has been disconnected, then remove mains leads from terminals (after drawing a diagram of the connections).
2. Remove motor from its cradle (if fitted) in preference to removing base securing bolts.
3. Mark end-covers in relation to stator casing.
4. Undo and remove the four long end-cover securing bolts, and ease off the cover at the pulley end.
5. Withdraw end-cover complete with shaft and rotor. This is best done downward to ensure that shaft spacing shims remain in position. Examine inside casing for any dislodged shims.

6. Examine centrifugal switch attached to the rotor. Clean, free, or replace as required. This is usually a simple and obvious operation consisting of removing two screws.

7. Gently ease off the other end-cover, bending back carefully to avoid straining the leads. Clean or replace switch as necessary.

SWITCH

To renew a switch a soldering iron is required to remove the leads, and it is advisable to use a few 'price-labels', which can be marked and attached to the leads to prevent confusion when replacing. Whilst the leads are disconnected, if the insulation is hard and brittle, it should be removed and the leads re-insulated with rubber sleeving, which in turn should be covered with protective sheathing. If the thermal overload (often incorporated in the terminal block) has been cutting on and off for a considerable or unknown period, it is advisable to change this at the same time.

Finally, remove any dust from the windings, give each bearing a soaking with oil, and reassemble, again taking care that all shaft shims are in position. If any shims are omitted, the shaft will have excessive 'end-play', *i.e.* the shaft can be moved horizontally more than is necessary for free rotation. This is sometimes the cause of contact failure in the centrifugal switch, and where this exists through normal wear, extra shims must be added.

REASSEMBLY

With the repair complete, reassemble the motor. Ensure marks are realigned correctly, and gently press the end-covers into position. Make sure that the shaft is free and can be turned by hand, then replace the long bolts and tighten up evenly. Check again that the shaft is free.

An apprentice should undertake the above under supervision in the workshop, but after a little experience it can quite easily be carried out on site. This results in a much more economical job, since the time of adapting a standby motor, and ultimately of returning, is avoided.

17

The Domestic Refrigerator

The refrigerator used in the home employs what is known as a 'sealed unit', and whilst this kind of unit is being made in ever-increasing sizes, it is the sealed unit of the domestic refrigerator with which we are concerned in this chapter.

OPEN UNIT

The 'open' or belt-driven unit has many disadvantages when applied to the small domestic refrigerator. It is somewhat clumsy and inclined to be noisy; moreover its many 'flared' joints, and its compressor shaft seal, make it prone to repeated gas leaks. In addition to which, the motor, being exposed to dust blown through by the condenser fan, also needs regular attention.

SEALED UNIT

The sealed unit overcomes these difficulties. In the first place the motor and compressor are contained in the same casing, with the compressor directly driven. It is therefore running faster than the open compressor, so can be made smaller whilst still doing the same amount of work. These factors result in a much more compact component. The need for a shaft seal has been removed, and, as the whole system is totally enclosed, with joints brazed, and gauge connections omitted, the risk of a gas leak is reduced to a minimum, and the possible intake of dirt and moisture eliminated.

REFRIGERANT CONTROL

The regulation of refrigerant flow is by capillary tube (*see* p. 39). This is cheap, easy to incorporate, and requires no adjustment once the correct length has been ascertained. It is also very suited to a small gas charge, providing the correct operating pressures when running, but reverting to an equalised or balanced pressure throughout the system when stopped. This has the effect of reducing the starting torque on the motor, allowing a split-phase type to be used, thus obviating the necessity of a capacitor.

CONDENSER

One other feature of the domestic refrigerator is the condenser, which is designed with a surface area large enough to achieve its purpose of liquefying the pressurised gas, without the necessity of a fan. These elements all make for economic production, resulting in a neat, quiet-running piece of machinery, which in the main gives years of trouble-free refrigeration.

Despite this, sealed units still break down, and whilst in most cases they are repairable, the manufacturers' five-year guarantee, plus a replacement scheme after this, causes domestic units to be changed rather than repaired. However, the majority of service calls on domestic refrigerators are not due to the sealed unit at all but to the controls. These are as follows:

CONTROLS

THERMOSTAT

The general construction of a thermostat has been described on p. 60 and the important point to remember when fitting it is to avoid manipulating the capillary more than necessary. Bear in mind that a capillary is a fine bore tube, so unwind it gently, and avoid sharp bends. Having unwound just sufficient to reach the phial clamp, ensure that it is firm and making good contact with the evaporator.

RELAY

The function of the relay is to start the motor. Obviously the centrifugal switch of the open motor could not be used in an enclosed motor. Quite apart from the insulation difficulties, the possibility of physical breakage would be too great a risk.

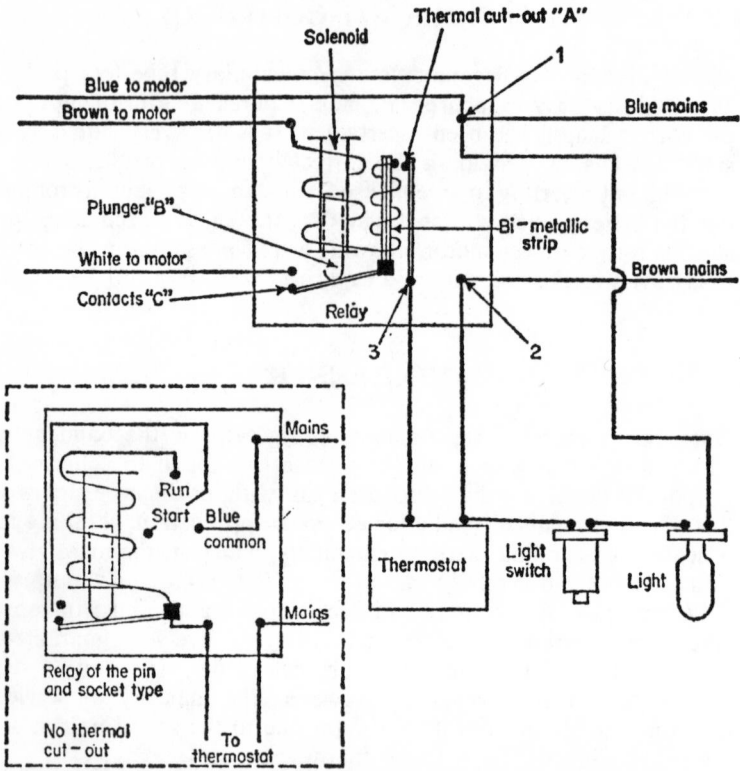

Fig. 45. Relay and wiring diagram of a domestic refrigerator. The relay performs the same function as a centrifugal switch, providing current to the starting windings initially, then cutting it off as soon as the motor starts.

Figure 45 shows the wiring diagram of a typical domestic refrigerator, including working details of the starting relay. The wiring is quite simple and self-explanatory. The mains lead enters the relay, the blue or 'neutral' lead passing straight through to the blue lead, or common terminal, of the motor. The brown, or 'live', lead is led up to the thermostat and thence back to the relay. It now passes via the thermal cut-out A (which in principle is identical with that of the open motor) through the solenoid and direct to the run winding, or brown terminal, of the motor, and the action is as follows:

As the current is switched on with the motor at rest, the current passing through the solenoid is at its maximum, creating a sufficiently strong magnetic field to lift plunger B, permitting contacts C to spring together. This allows current to the start or white terminal of the motor, energising the start windings, and the motor starts. Once running, the current declines as a result of which the magnetic field is weakened, and is no longer capable of supporting plunger B. It thereupon drops, breaking contacts C, leaving only the run winding in circuit. It can be seen that for the plunger to function properly, the relay must be fitted in one position only, and the 'top' is so marked to ensure this.

Relays are designed to suit individual units, and should not be interchanged with other machines. If a thermal overload is incorporated as shown in the diagram, it would be most unlikely to match the current consumption of another unit, resulting either in inadequate overload protection, or persistent cutting out.

OVERLOAD PROTECTOR

The overload is more commonly fitted as a separate item, as illustrated in Fig. 46. The thermal disc, normally arced and making contact as shown, straightens on being heated, thus causing the contacts to break. The disc may become heated to this extent either by the conducted heat of the compressor, or by the small element contained within the overload, which overheats if subjected to a current in excess of that rated for the motor. These overloads are very sensitive, and again it is important to replace this item with one rated for the motor.

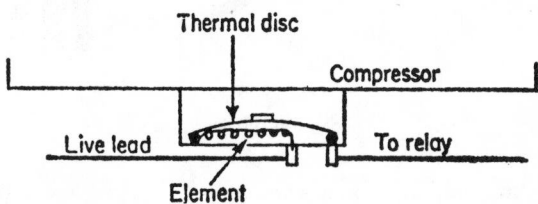

Fig. 46.   Overload protector. The metal disc breaks contact either by conducted heat of the compressor dome, or by excess current through the heater.

ABSORPTION REFRIGERATORS

Figure 47 is a simplified sketch illustrating the system which is achieved within a circuit containing a constant pressure, and without the aid of moving parts.

The three ingredients of the unit are water, ammonia and hydrogen, and the following basic principles are utilised:

1. The capacity of water to absorb large quantities of ammonia vapour very readily.

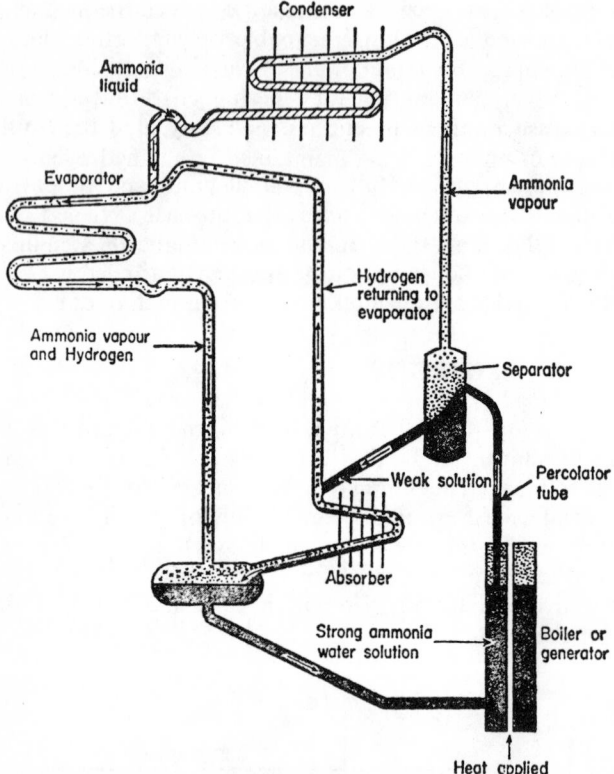

Fig. 47.   Absorption refrigeration system.

2. The fact that ammonia vapour will be given off from this solution if heat is applied.
3. Dalton's law of partial pressures which shows that in a space occupied by a mixture of a vapour and a gas which do not react chemically with each other, each exerts the pressure which it would produce if it alone occupied the space, and the total pressure is equal to the sum of these two pressures.

With a source of heat, which can be from electricity, gas or paraffin, the cycle is as follows:

Heat applied to the generator or boiler causes the strong solution of water and ammonia to be passed upward through the percolator tube into the separator, where ammonia vapour continues to rise to the condenser, while water drains back to the absorber. Ammonia vapour cooled in the condenser at a pressure of about 13·33 bar (200 lb)/in² liquefies, and drains by gravity into the evaporator in which hydrogen gas is retained. It is here that the mixture of ammonia and hydrogen gases takes place, and where Dalton's law applies. Liquid ammonia evaporates into a gas, taking up latent heat from the surrounding metal and thereby achieving refrigeration.

At this point ammonia pressure is approximately 1 bar (15 lb)/in^2, the balance of pressure being maintained by the hydrogen gas. The resultant ammonia/hydrogen mixture, being heavier than hydrogen, falls to the absorber where it meets weak solution coming from the separator. Such is the eagerness of water to absorb ammonia that the low pressure of ammonia is maintained, while hydrogen, being insoluble in water, separates and rises through the absorber piping and returns to the evaporator. The strong solution in the absorber now drains back to the boiler for the cycle to be repeated.

The unit is constructed of welded and sealed steel piping, and a serviceman could not repair on site an absorption unit which had a leak or had failed owing to some internal fault. The unit is made in a factory and must be replaced in the field by a new unit. There are, of course, problems arising from causes other than unit failure which the serviceman can cure.

A requirement of all absorption units is that they be level, otherwise internal liquid flow may be retarded or even curtailed altogether. Also a clear flow of cool air over the condenser is essential. A few causes of refrigeration failure are listed as follows:

ELECTRICALLY HEATED

Possible fault	Remedy
1. No power to machine	Check power socket, plug top, and supply leads
2. Thermostat at fault	Change thermostat
3. Heater element failed	Replace element

GAS-HEATED

1. No gas to machine	Check gas supply
2. Gas jet choked	Remove and clean
3. Constant pressure valve failed	Replace pressure valve
4. Flue choked	Remove jet assembly. Plug bottom of flue with cork, and fill flue pipe with boiling water. Drain into a bowl. Repeat this process until the scale is softened and can be removed with a suitable wire brush.

If all the above have been checked and found in order, and still no refrigeration takes place, it is worth turning the machine upside down. Do this carefully, after allowing the unit to cool, by laying the cabinet first on its back, and then on to its top, using a protective blanket to avoid possible damage to the cabinet. Repeat this procedure several times, then restore the power supply to the unit, after allowing time for the liquid and gas inside to settle.

An absorption unit operates more satisfactorily when the flow is continuous, and gas-heated machines, where the thermostat regulates the gas supply to the jet, are generally more suited to the system than a sudden on/off as given by an electric heater element. The thermostat of an electric machine must have a close differential to enable operation to be restarted before defrosting of the evaporator takes place. Some machines overcome this slight disadvantage by using a 'two-leg' element, one leg of which is on all the time, the other being brought in periodically by the thermostat.

ADVANTAGES OF THE ABSORPTION SYSTEM
 a. Silent operation.
 b. No moving parts.
 c. Refrigeration made possible where electric supply not available.

DISADVANTAGES
 a. Rather slower freezing than compression machines.
 b. Slightly more expensive to operate.

SERVICING NOTES

The diagnosing of faults depends on observation, knowledge and experience. The two latter naturally take time to acquire, and observation is of little value unless the correct operation of a machine is known, so the following list is provided:

NORMAL OPERATION OF A DOMESTIC REFRIGERATOR
 1. Sealed unit audible, but not obtrusively noisy.
 2. Running time between three and seven minutes 'on', and between six and fourteen minutes 'off'. Ratio approximately one 'on' to two 'off'.
 3. Cut-out time from initial start 15 to 45 minutes.
 4. Evaporator temperature between $-15°C$ (5°F) and $-10°C$ (15°F) with thermostat at 'normal'.
 5. With thermostat at 'warmest', cut-out temperature between $-10°C$ (15°F) and $-7°C$ (20°F).
 6. Cabinet temperature between 4 and 7°C (40 and 45°F). (Not itself adjustable but dependent on the evaporator temperature.)
 7. Stopping and starting clean and positive, with no hesitation or excessive vibration.

8. Gasket fitting neatly round the door with no gaps between it and the cabinet.
9. Interior light going out when the door is shut.

DEFROSTING

Room temperature, or water recently put into the ice-trays, will vary conditions 2, 3, 4 and 5, and allowances must be made in these circumstances. The evaporator should not be over-frosted, and advice to customers should be to defrost the evaporator once a week, and to clean the cabinet and shelves at the same time. Defrosting simply entails switching off and allowing the frost formation to thaw and drain into the drip-tray, after which excess moisture is wiped off with a clean cloth.

SEALED UNIT TESTER

Illustrated in Fig. 48, this useful piece of equipment is in effect a manually operated relay and is used as follows:

a. Disconnect leads on sealed motor terminals.
b. Connect appropriate leads from the tester to motor terminals.

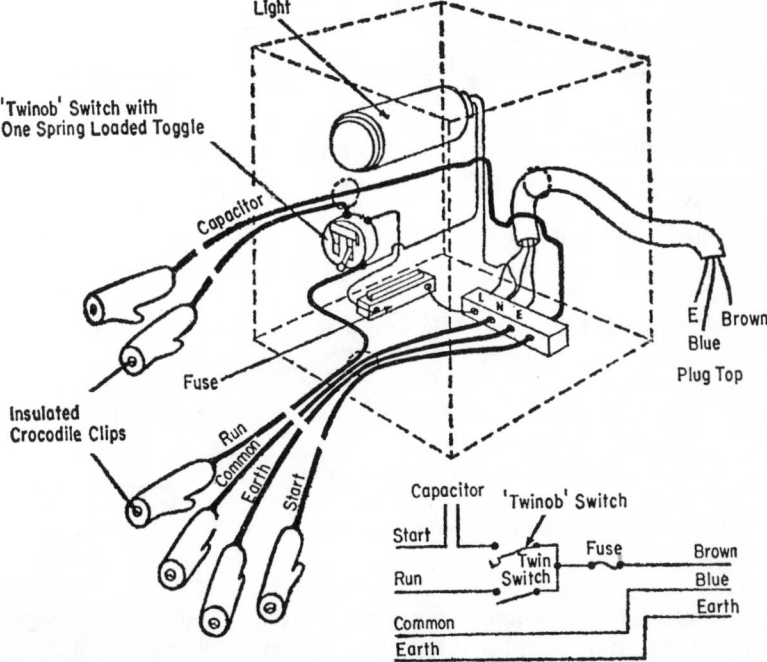

Fig. 48. Sealed unit tester. This instrument allows a hermetic compressor to be started manually, and thus tested before fitting a new relay.

c. Connect the two capacitor leads together (assuming no capacitor is employed).
d. Connect to mains and switch on.
e. Press the switch of the tester.

If the motor is in good order it will start, but if it does not, switch the tester off after a period of not longer than four seconds. Testers, sometimes known as 'test cords' or 'flash testers', are available from most manufacturers.

COMMON COMPLAINTS

The following are common complaints associated with a domestic refrigerator:

COMPLAINT
Motor running but machine not freezing.

Possible Fault
A. System short of refrigerant.
B. Capillary blocked.
C. Compressor not pumping.

Comment
The complaint can be confirmed only by observation.

Remedy
Change the unit in either case.

Complaint
Unit runs too much.

Possible Fault
Partial conditions of A, B or C.
D. Excessively high ambient.
E. Retarding air flow through the condenser.
F. Gap in door seal.
G. Thermostat set on 'coldest'.

Test
Check the evaporator temperature preferably with the ice-trays removed. Refer to 'Normal Operations' numbers 4 and 5, and note too if the evaporator is frosted all the way round. Sometimes when the unit is losing refrigerant (which it may do very slowly over

several months) a condition can be reached where partial freezing gives a low temperature reading in one part of the evaporator, whilst the temperature by the thermostat phial clamp is above −10°C (15°F). (It is no use changing the thermostat to cure this condition.)

Remedy
According to the observations. For A, B and C, change the unit.

COMPLAINT
Temperature too cold.

Possible Fault
G. Thermostat set on 'coldest'.
H. Thermostat stuck in closed position.
I. Thermostat phial not securely clamped.
J. Ambient temperature too low (below 15°C (60°F)).
K. Short circuit on thermostat lead.
L. Thermostat out of adjustment.

Comment
This complaint may also be linked with that of 'unit runs too much'.

Remedy
Adjust or change thermostat if shown to be at fault.

COMPLAINT
Not cold enough.

Possible Fault
L. Thermostat out of adjustment.
M. Thermostat set on 'warmest'.
N. Evaporator in need of defrosting.

Remedy
Adjust or change thermostat if shown to be at fault.

COMPLAINT
Motor not running.

Possible Fault
O. Fuse blown.
P. Motor at fault.
Q. Short circuit, or open circuit in the wiring of controls.

Test

The 'surge' load on initial start-up of a refrigerator may be six times that of the running load, consequently a low amperage fuse may blow for this reason only. A slightly higher rated fuse will cure this. On the other hand, a blown fuse is generally a warning of a more serious fault, and it is advisable to check carefully through the circuit, looking for signs of charring at obvious places such as the relay, overload, thermostat or perished wiring. A 'Megger' (*see* Glossary) is a valuable instrument on these occasions, but on the simple circuit of a refrigerator, observation is usually adequate. If no evidence of a short is found, use the sealed-unit tester as described.

If the motor is now found to be in order, then disconnect the thermostat at the relay, and recheck using a piece of insulated flex in place of the thermostat circuit. If the fault is still not located, try another relay or overload.

If, after repairing the fuse, it does not blow immediately but the motor fails to start, make sure the thermostat is switched on, then check the circuit using the test lead as described on p. 74. For example, if a light is obtainable on points 1 and 2 of Fig. 45, but not on points 1 and 3, an 'open' circuit is confirmed on the thermostat or thermostat wiring. Test in this way until the break is located.

Remedy

Replace item found to be at fault. If motor at fault, change the unit.

SERVICE INFORMATION

It must be appreciated that a large number of different cabinets are in daily use, representing various models over the years, and it is impossible to give precise details to cover all of them. However, service books are obtainable from manufacturers, which provide essential information for changing a unit, or replacing a thermostat, and also include part numbers for spare parts. In addition, special features such as press-button defrosting, or adjustable door latches, make specific instructions for one type of machine indispensable. All the same it is unwise to produce this in front of customers, who tend to assume reference to a book implies incompetence; therefore read information like this quietly in the van.

COMPLETION

Finally, remember that domestic refrigerator service entails entering the housewife's kitchen, so be especially clean and presentable.

Be polite but not over-familiar. Move the refrigerator gently to avoid possible scratches on the tiles or linoleum, and when complete (with all relevant numbers recorded) replace the machine in exactly the same position, remove any finger prints from the cabinet and debris or marks from the floor. Now complete the necessary paperwork, and obtain the customer's signature.

18
Humidity and Defrosting

FROST

On pp. 78 and 97 mention has been made of 'defrosting'. The frost formation in a domestic refrigerator will be familiar to most readers, and this steady frost build-up on the evaporator extends throughout all applications of refrigeration. This often causes considerable problems, and from time to time it must be cleared to ensure efficient cooling, but before proceeding further let us consider the reason for frosting taking place, and why it is unavoidable.

ATMOSPHERE

The atmosphere contains a quantity of water particles in suspension, derived from evaporation of water surfaces, which takes place at all temperatures. The amount of water vapour held in suspension varies, but is limited by the capacity of the air itself, which at any given temperature can only hold a certain amount. At this point it is said to be saturated—rather like a soaked piece of blotting paper—and any reduction in temperature at saturation will cause excess vapour to be condensed back into water. If the temperature is increased, however, a greater quantity of aqueous vapour can be held.

This vapour in the air is known as the humidity, and it is constantly fluctuating and varying throughout the world. India, for example, has a predominantly humid atmosphere, whilst the desert areas of the Middle East have a very dry atmosphere. Extremes of either are undesirable from a comfort point of view, and control of humidity plays an important role in air-conditioning.

RH

The humidity or 'wetness' of the atmosphere is measured by the percentage of water actually held in suspension, at any given temperature, against the total quantity which could be held at saturation. This is known as the Relative Humidity, more commonly referred to as RH.

It must be emphasised that the *quantity* of aqueous vapour present depends on the temperature. Thus more vapour is present on a hot day with a RH of say 70, than on a cold day with a RH of 70.

HYGROMETER

This is an instrument which is designed to provide visual indication of the prevailing RH by means of a moving needle and dial.

DEW POINT

As we have seen, at saturation point any reduction in temperature will result in the aqueous vapour condensing back into water. If the atmosphere has a RH of 70, however, any cooling will not immediately cause condensation, but will have the effect of raising the RH until saturation point is reached. Further reduction at this stage will cause condensation, and this is known as the 'dew point'. Evidence of this feature is found in the evening, when the temperature drop causes water vapour to be precipitated in the form of dew, or in cold weather, as frost.

FROST BUILD-UP

It should be clear now, therefore, that an evaporator being mechanically cooled below the dew point of the surrounding air, will result in moisture being deposited on it. Since the evaporator temperature is generally below freezing point, this moisture forms as frost, and since doors are normally being opened and shut, allowing more moisture-laden air to replace the dried air, the frost formation gradually builds up.

This frost build-up is inevitable, but by no means desirable. Cooling efficiency of the evaporator surfaces is reduced as the frost thickness increases, fins of blower-type coils quickly become completely blocked, rendering the fan ineffective, and the fan blade may become fouled causing the fan motor to burn out.

It is to be remembered that the evaporator temperature is at least −8°C (15°F) below that of the coldroom temperature (TD: *see* p. 36), thus even with a coldroom temperature of 5°C (40°F) the evaporator temperature will be say −4°C (25°F), and frosting will take place. However, during the 'off' cycle, when the evaporator fan is usually kept running, the comparatively warm air will melt the frost, so preventing a slow build-up. Any coldroom temperature below 5°C (40°F), say 2°C (35°F), will mean an evaporator temperature of −7°C (20°F) and now the 'off' cycle will not clear the evaporator. Consequently, to maintain a constant temperature below 5°C (40°F), some artificial means of clearing frost formation is necessary. For temperatures of 3°C (37°F) this need only be a time switch set to stop the plant during the night for a period of about two hours, leaving the evaporator fan running. For temperatures below this more positive means must be used, and it is for this reason that various methods of artificial defrosting have been developed.

TIME SWITCH

One item which is common to all defrost systems is the time switch. It consists basically of an electric clock, incorporating a mechanism which makes and breaks electric contacts at predetermined intervals. Time switches vary in design according to the

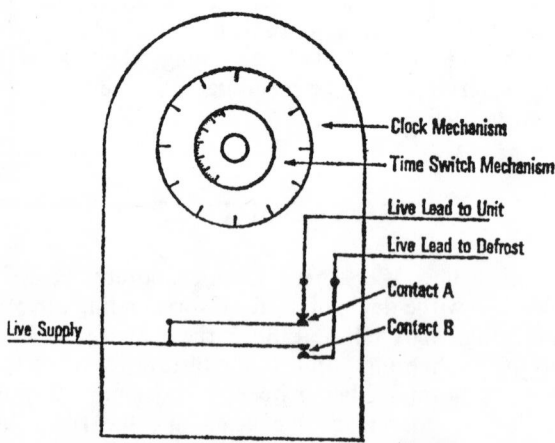

Fig. 49. Time switch. This component is widely used to perform defrosting automatically at regular intervals. When one contact is made the other is simultaneously broken.

manufacturer, and may possess special features relative to requirements. The operative parts are illustrated in Fig. 49. It can be seen that current is sent either to contact A or contact B, for a period as pre-set and timed by the clock. As many as twelve operations per day can be provided, but three or four defrost actions are normally sufficient.

Students are advised to study literature relating to the various types of time switches in use.

ARTIFICIAL DEFROSTING

Artificial defrosting on commercial refrigeration plant is achieved mainly by water, electricity, hot-gas or 'reverse cycle'. Each has its advantages and disadvantages detailed as follows:

WATER DEFROST

In this system the evaporator is constructed with a perforated tray above the fins (*see* Fig. 50), which is connected to a water supply as shown. Beneath the fins is a drip tray, connected to a permanent drain pipe.

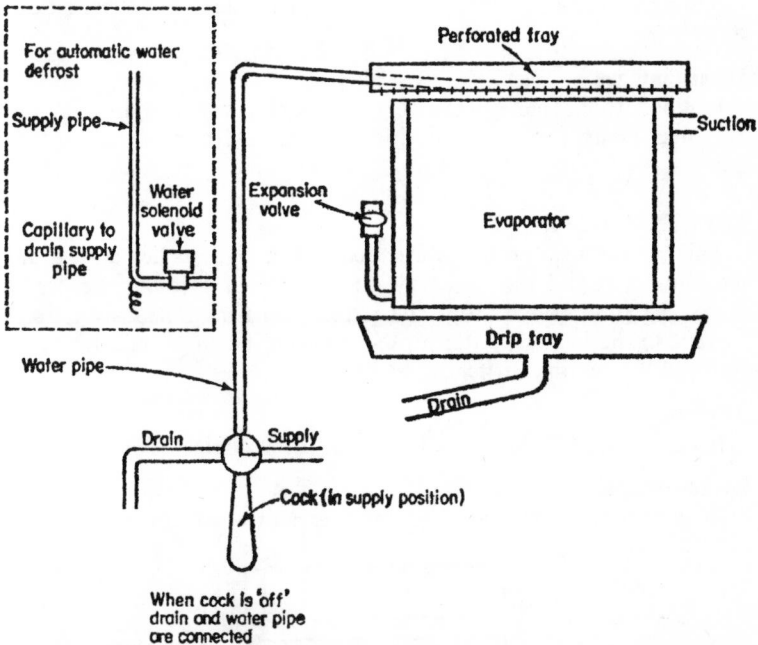

Fig. 50. Water defrosting: a simple method of removing excess ice formation from an evaporator. Note the special two-way cock required to ensure the supply pipe is drained when not in use.

The principle is quite simple. The plant is switched off, the fans stopped, and water to the tray turned on. The perforations spread the water rather in the manner of a watering can, thawing and washing excess frost from the fins into the drip tray. When defrosting is complete, the water is turned off, and after a period of at least ten minutes (allowing water to drain clear) the plant is switched on again.

It is important that the drain pipe be of generous proportions, and that it has a rapid fall, with few bends. Also the supply pipe must be fitted with some means of draining after use, such as the two-way cock shown in the sketch. This also prevents water entering the tray during freezing, should the supply tap leak or not be shut off properly. Without these precautions trouble with water freezing and blocking of pipes is bound to occur. Automatic action can be incorporated by using a solenoid valve in place of the water cock with a capillary for draining the supply pipes, as shown in Fig. 50. In this case a delayed restart action time switch is essential.

Advantages
Simple to install. Cheap to use. Is independent of refrigeration machinery.

Disadvantages
Inclined to be messy. Danger of blocking through inadequate drainage. Danger of water spilling into the coldroom, or on to fan motors.

ELECTRIC DEFROST
This type of evaporator, illustrated in Fig. 51, has heaters within the finning, round the drip tray, and round or through the drain pipe. The time switch stops the plant, and simultaneously passes current to the heaters. After a pre-set period normal conditions are restored by the time switch.

Advantages
Easy to install. Does not interfere with the refrigeration system.

Disadvantages
Evaporators more expensive. Fairly expensive to run.

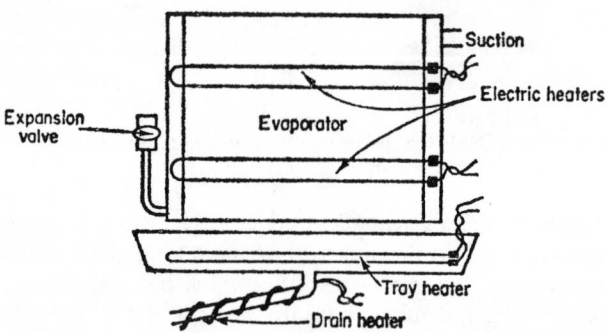

Fig. 51. Electric defrosting. Electric heaters in the drip tray, drain, and within the fins, melt the excess frost. Automatic action is obtained by the use of a time switch.

HOT-GAS DEFROST

Figure 52 shows a hot-gas evaporator. It has extra piping attached to the drip tray, which then leads to the 'T' inserted after the expansion valve. This piping is connected to the compressor discharge, and isolated by a solenoid valve (*see also* p. 78). When the time switch operates, it stops the evaporator fan, energises the solenoid valve, and the plant remains running. The hot-gas bound for the condenser is now by-passed and directed via the extra piping and 'T' to the evaporator.

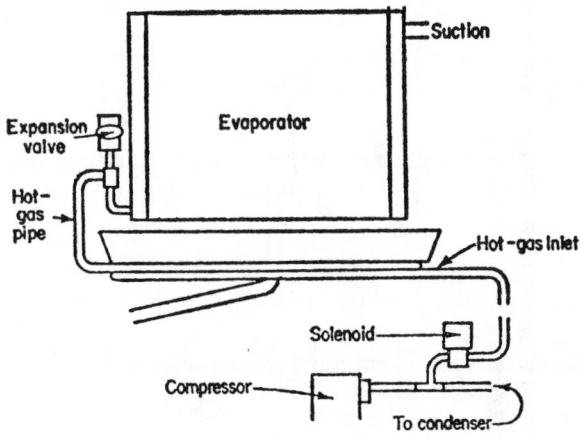

Fig. 52.   Hot-gas defrost. Hot gas from the compressor is led direct to the drip tray and evaporator, by-passing the expansion valve.

The result of this (apart from defrosting the coil) is to liquefy the refrigerant in the evaporator, and some means must be incorporated to prevent this liquid being drawn back into the compressor. Manufacturers who use hot-gas defrosting have their own particular methods, and the apprentice is advised to study any hot-gas system to see which is being used. The important point to remember is that liquid must not be allowed back to the compressor before evaporation.

Advantages

Relatively cheap to install. Economic in operation. Is more easily adaptable to evaporators already in use, but which have no existing defrosting means.

Disadvantages

Extra pipe line required. Uses the refrigeration plant. Possibility of leaking solenoid. Not always effective in low ambients.

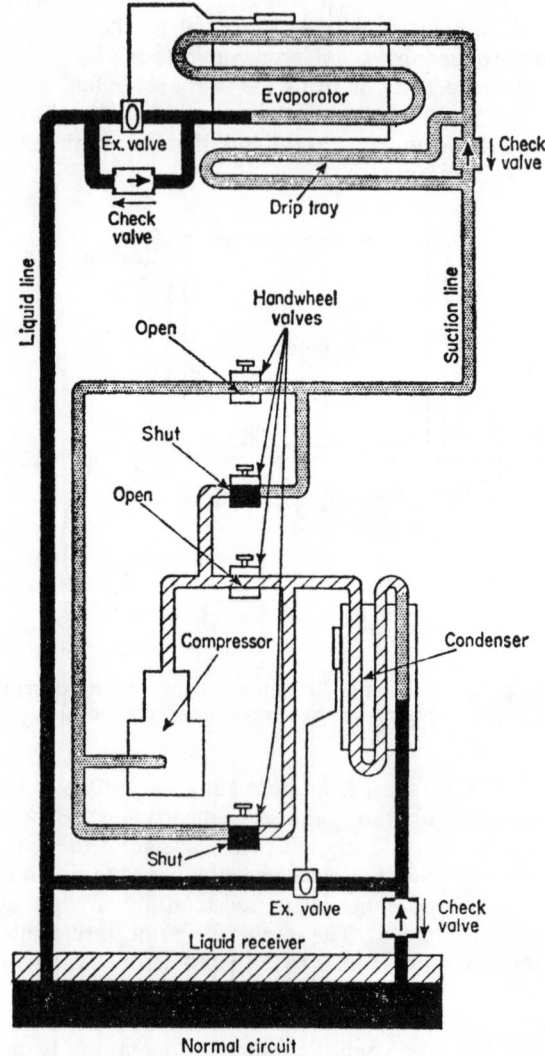

Fig. 53. Reverse cycle defrosting. The normal circuit of a refrigeration cycle can be reversed by means of handwheel valves and non-return valves as shown. Thus the evaporator temporarily becomes the condenser and the condenser the evaporator.

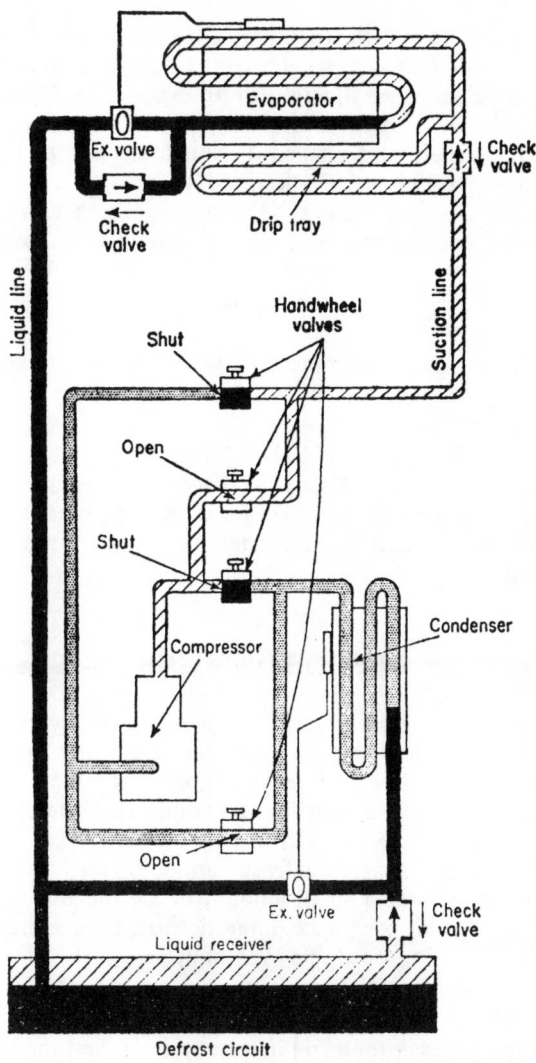

Fig. 53.

REVERSE CYCLE

'Reverse cycle' means that the function of the evaporator is made to perform that of a condenser (similar to the hot-gas system), and the condenser to perform the function of an evaporator. This may be achieved by the use of hand-wheels and non-return valves suitably placed in the circuit, and opened or shut manually, or by the use of a special 'reversing valve', designed to provide automatic action. The circuit is shown in Fig. 53. Two expansion valves are required, and study of the two diagrams will show how 'reversal' is accomplished.

Advantages

Very effective defrost. Economical in use. May be adapted to existing plant.

Disadvantages

More complicated to install. Uses the refrigeration system.

Note: Use of the refrigeration system for defrosting is considered to be a disadvantage, because a failure may impair the refrigerating operation as well.

FAULTS ON DEFROSTING SYSTEMS

Failure of a defrost system is clearly evident by the evaporator being encased in ice, although quite often the complaint from the customer is 'machine not freezing' or 'temperature going up'.

Always examine the time switch first to ensure that the clock is going and the contacts 'making'. On many occasions excessive ice formation is due to too short a defrost period, or an insufficient number of periods. Even if a small amount of ice remains after a defrost, it will grow with each succeeding defrost, so a minute or two short is enough to cause trouble eventually.

OTHER FAULTS

The main causes of other faults on defrost systems, and the things to look for, are:

Water

Ice blocking the water pipes. Water not reaching all parts of the evaporator.

Electric

Heaters burnt out.

Hot-Gas

Failure of solenoid valve.

Reverse-Cycle

Reversing valve not operating. Expansion valve at fault. Blockage in system. Non-return valve stuck.

19
Insulation and Load Calculation

This chapter is intended to be only an introduction to the factors involved concerning cabinet insulation and plant load. It is perhaps not strictly necessary to know these facts in diagnosing a mechanical fault; on the other hand, for those wishing to advance in refrigeration, and for those wishing to be fully conversant with their trade, they will be of value. Moreover there are occasions when the plant is being overloaded or the insulation is inadequate, and it is of no use to blame a unit for failing to cope with a load beyond its designed capacity. Some knowledge of insulation principles and load calculation is therefore essential.

INSULATION

On p. 30 it was shown that heat will always flow from a hot body to a cold one. Mechanical refrigeration creates a 'cold' area, which the surrounding heat tries to reach, so a barrier must be placed between the hot and cold areas to retard this flow of heat. This 'barrier' is called insulation, the effectiveness of which depends on the material being used and its thickness.

It would be possible, for example, to insulate a coldroom to the extent that practically no heat penetrated at all, but this would be both costly and cumbersome beyond necessity. Thickness of insulation is therefore a compromise between cost, effectiveness and practicability. If it is too thin, then heat penetration will impose an undue strain on the plant, and also condensation will form on the cabinet exterior. Approximately 25 mm for every 7°C temperature difference is considered suitable, although it depends on the 'k factor' of the insulant being used.

MOISTURE

In conjunction with heat penetration, there is also the risk of moisture permeation. The atmosphere surrounding a cabinet normally contains a greater vapour pressure than that of the air inside, as a result of which precautions have to be made to eliminate the ingress of moisture, particularly through brick walls. Should insulation become sodden, through ineffective vapour sealing, the insulating quality of the material is destroyed.

Insulation of a coldroom is made up as illustrated in Fig. 54, with vapour seal on the outer wall, usually in the form of a bituminous compound applied with a brush; insulation blocks, staggered to ensure air-tightness, sealed with a suitable adhesive; and an inner lining normally of asbestos, the facing side being glazed.

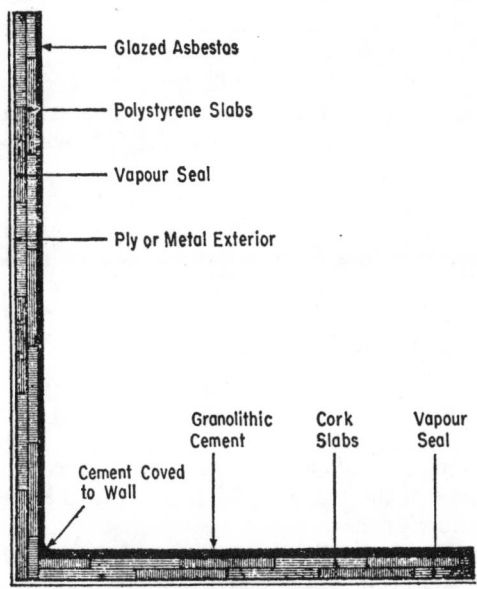

Fig. 54. Coldroom insulation, showing a section of the floor and side of a standard coldroom.

INSULATING MATERIAL

The most efficient heat barrier would be a vacuum, but for general purposes this is impracticable. Air is an excellent insulant in itself but quickly transfers heat by convection. The most effective insulants are those which are cellular, in other words substances containing minute air pockets.

Apart from the obvious requirement of low heat conductivity, a good insulant must be light, easy to handle, robust (particularly for floors), cheap, vermin-proof and durable. No material can claim to possess all these qualities, but cork satisfies most of them and has for many years been the leading insulant. Modern plastic foam insulants have been produced, however, which in some respects are superior to cork.

HEAT CONDUCTIVITY

THE 'k' FACTOR (See Table 19.1, p. 123)

To assist in selecting an insulant, and to calculate the thickness required, a standard of heat conductivity has been devised known as the 'k factor'. This is the amount of heat, in Btu per hour, which will pass through a section of the material 1 ft sq and 1 in thick, when subjected to a temperature difference of 1°F.

In SI units, heat flow is measured in watts for a 1 m cube of the material and a temperature difference of 1°C (1 watt = 1 joule per second).

Table 19.1 provides a list of insulants most commonly used for coldrooms, giving the 'k factor' in each case, in both Imperial and SI units, together with a few other items used in building work. It can be seen that bricks are poor insulants, and in calculating heat penetration can be ignored.

The formulae for the rate of heat (Q) is given by

$$Q = \frac{kAT}{d}$$

where k = thermal conductivity obtained from the chart

A = area of the material through which the heat is flowing (in ft^2 or m^2)

T = temperature difference between the two faces of the material (in °F or °C)

d = thickness of the material (in inches or metres).

It is of course essential to select k in the right units from the table and never to try to work in a mixture of units which are not the same as those of k.

Imperial Units

Example A

Suppose we have the following facts concerning a butcher's coldroom which is required to operate at a temperature of 35°F.

External area of coldroom = 400 ft^2 (this includes ceiling and floor)
Cork insulation, 'k factor' = 0·3
Temperature difference = 50 (taking ambient of 85°F)
Thickness of insulation = 4 in

Rate of heat flow
$$= \frac{k \times A \times T}{d} \text{ Btu/h}$$

$$= \frac{0·3 \times 400 \times 50}{4}$$

$$= 1\ 500 \text{ Btu/h}$$

Example B

A coldroom is to operate at a temperature of 20°F, and the insulation is to be 5 in of foam plastic. The exterior dimensions of the room are 10 ft × 9 ft × 8 ft. In an ambient of 85°F what is the calculated heat penetration?

Area of walls, floor and ceiling $= (10 + 10 + 9 + 9) \times 8$
$$+ 2(10 \times 9)$$
$$= 38 \times 8 + 2 \times 90$$
$$= 484 \text{ ft}^2$$

$$k \text{ (from table)} = 0·24$$
$$T = 85 - 20 = 65$$
$$d = 5 \text{ in}$$

Rate of heat flow $= \dfrac{k \times A \times T}{d} \text{ Btu/h}$

$$= \frac{0·24 \times 484 \times 65}{5}$$

$$= 1\ 510 \text{ Btu/h}$$

SI Units

Example C

Coldroom to operate at 2°C. External area 40 m^2, k factor 0·04, temperature difference 30°C, insulation thickness 0·1 m.

$$Q_1 = \frac{0·04 \times 40 \times 30}{0·1} = 480 \text{ watts}$$

$$= 0·48 \text{ kW}$$

Example D

Coldroom at $-6°C$, ambient $30°C$, insulation 0.15 m thick, $k = 0.035$, external dimensions 3 m $\times$ 3 m $\times$ 2.5 m. What is the heat load?

$$\text{Areas } 2.5(3 + 3 + 3 + 3) + 2(3 \times 3)$$
$$= \qquad 30 \qquad + \quad 18$$
$$= 48 \text{ m}^2$$

Temperature difference $= (30 - (-6)) = 36°C$

$$Q = \frac{0.035 \times 48 \times 36}{0.15}$$

$$= 404 \text{ W}$$
$$= 0.404 \text{ kW}$$

FLOOR

A more accurate figure may be obtained by calculating the floor separately and using a ground temperature of $12°C$ ($55°F$), but on a small coldroom the difference is insignificant.

FANS

The heat generated by evaporator fans, lights, and heat gains through door opening must also be taken into account. When working in SI units the fan and lighting loads can be taken directly in watts. For Imperial calculations the values in Btu/h may either be obtained from the relevant literature or calculated using the following facts:

$$1 \text{ watt} = 3.41 \text{ Btu/h}$$
$$1 \text{ hp} = 2\,540 \text{ Btu/h}$$

Where the service load (door opening) is light, a figure of 10% of the heat penetration load is considered adequate. Where the service load is heavy then 20% should be taken. Special circumstances must be dealt with accordingly.

PRODUCT LOAD

In addition to heat penetration, refrigeration plant must have sufficient power to extract heat from the produce being stored. This is known as the 'product load'. If a coldroom receives goods which are already down to their storage temperature, frozen foods for example, no product load is involved, but if meat is being delivered in an ambient temperature of perhaps $18°C$ ($65°F$), and put into a coldroom to be reduced to say $2°C$ ($35°F$), then the product load is considerable. To calculate the product load an understanding of specific heat is necessary.

Referring back to p. 18 we have seen that it requires 1 Btu to be given to, or taken from, 1 lb of water to raise or lower its temperature by 1°F. In general, metals in particular require less heat than 1 Btu to achieve this. The temperature of 1 lb of aluminium, for example, can be raised 1°F by the application of one-fifth of a Btu or, as a decimal, 0·2, and this ratio is the specific heat.

SPECIFIC HEAT

DEFINITION
Specific heat in the Imperial system is defined as:

> *The ratio of the heat required to change the temperature of a substance, to the amount of heat required to change an equal mass of water through the same number of degrees.*

SPECIFIC HEAT CAPACITY
Because the SI unit of energy (the kilojoule) is not directly defined in terms of water it is usual to refer to 'specific heat capacity' (SHC) of materials, and this is expressed in kJ/kg°C. (Strictly speaking, the 'Imperial' specific heat also has units—Btu/lb°F—which must be used if the various formulae are to become dimensionally consistent.)

TABLES 19.3–19.4 (*See* pp. 126–127)
These give useful information relating to commodities stored in a refrigerated condition, including the specific heat of the item, in both Imperial and SI units.

Imperial Units
From the Imperial table we find that the specific heat of fresh lean beef is 0·77. If a butcher puts into his coldroom 1 cwt of fresh lean beef, which has been delivered to him at a temperature of 65°F, the quantity of heat to be extracted to reduce it to a temperature of 35°F is as follows:

Heat to be extracted = Mass of meat in lb × specific heat × TD
$$= 112 \text{ lb} \times 0·77 \times 30 \, (65 - 35)$$
$$= 2\,587 \text{ Btu}$$

Assuming that the butcher is satisfied with this temperature being reached in a period of, say, four hours, then:

$$\text{Btu to be extracted per hour} = \frac{2\,587}{4} = 647 \text{ Btu/h}$$

If we now link this product load with the heat load as demonstrated with the coldroom in example A, the total calculated load is as follows:

Heat penetration load	1 500 Btu/h
Service load	300
Fan load 75 W × 3·41	256
Product load	647
	2 703 Btu/h

Taking this example a step further, since the coldroom is to operate at a temperature of 35°F, the evaporating temperature will be 20°F (TD, p. 36); thus a unit with a capacity to extract 2 703 Btu/h, when evaporating at 20°F, is required to cope with the calculated load. The capacity of condensing units is obtainable from manufacturers' literature.

The heat having been considered so far is sensible heat (heat which can be measured by a thermometer, p. 16), but when the product temperature is to be brought down below its freezing point, latent heat of fusion must be included (heat which changes the state but not the temperature of the substance). This figure is considerable and makes a vast difference to the product load.

From Table 19.3, using Imperial figures, the latent heat of fresh lean beef is seen to be 100 Btu/lb. It is also seen that its freezing point is 29°F, and that there is a new specific heat (0·4) after the meat is frozen and still further reduced in temperature.

To reduce the temperature of any substance from above its freezing point to below its freezing point, three separate calculations have to be made:

Heat to be extracted above freezing point = Mass × sp. ht × TD
Latent heat to be extracted at freezing   = Mass × latent heat
Heat to be extracted below freezing point = Mass × sp. ht × TD

Example E

How much heat must be extracted from 1 cwt of fresh lean beef, to reduce its temperature from 65°F to 24°F?

Product load:

Sensible heat above freezing = *M.S.T.* = 112 × 0·77 × 36 (65 − 29)
$\qquad\qquad\qquad\qquad\qquad\qquad\quad$ = 3 104 Btu
Latent heat at freezing   = *M.L.*   = 112 × 100
$\qquad\qquad\qquad\qquad\qquad\qquad\quad$ = 11 200 Btu
Sensible heat below freezing = *M.S.T.* = 112 × 0·4 × 5 (29 − 24)
$\qquad\qquad\qquad\qquad\qquad\qquad\quad$ = 224 Btu

Total load =	3 104
	11 200
	224
	14 528 Btu

The extraction of latent heat represents by far the major part of product load, as can be seen, but the product load (not the heat penetration load) can be divided by the number of hours permitted for cooling.

Example F

The coldroom, as given in example B, is to be used to store half a ton of veal, which is delivered twice weekly at a temperature not exceeding 45°F. The veal is to be reduced to a temperature of 20°F in a time not exceeding 12 hours. What capacity of condensing unit is required? (Take details of specific heats above and below freezing, and latent heat, from Table 19.3 (viz. specific heat above freezing = 0·71; specific heat below freezing = 0·39; latent heat = 91; freezing point = 29).)

Product load:

Above freezing = M.S.T.		
= 1 120 × 0·71 × 16 (45 − 29) =	12 730 Btu
At freezing = M.L.		
= 1 120 × 91	= 101 900 Btu
Below freezing = M.S.T.		
= 1 120 × 0·39 × 9 (29 − 20) =	3 931 Btu

$$\text{Total} = 118\ 561 \text{ Btu}$$

Over a 12 hour period	=	9 880 Btu/h
From example B heat load	=	1 510
Service load 10%	=	151
Fan heat	=	1 000

$$\text{Answer} = 12\ 541 \text{ Btu/h}$$

The load in example F is an erratic one, in that only twice a week is the load at maximum, and then only for a period of 12 hours. For the remainder of the time a plant capable of extracting over 12 000 Btu/h would be more than adequate, and in this case two plants might be specified, one of which would be switched off other than at peak load periods.

Where the load is more constant, it is better to find the total load on a 24 hour basis, and then to divide this figure by the daily running time required in hours.

Example G

The total calculated load for a plant is found to be 96 000 Btu in 24 hours. What capacity must the plant be to give a running time of 16 hours?

$$= \frac{96\ 000}{16}$$

$$= 6\ 000 \text{ Btu/h}$$

The above examples are relatively simple, and apply to fairly small, standard type plants, but they are considered sufficient to give an apprentice an insight into the subject of insulation and load calculation.

SI Units

Example H

How much heat has to be extracted from 50 kg of fresh lean beef to reduce its temperature from 18°C to 2°C in 4 h?
SHC = 3·22 kJ/kg°C.

$$\text{Heat to be extracted} = \text{Mass} \times \text{SHC} \times \text{TD}$$
$$= 50 \times 3\cdot22 \times 16$$
$$= 2\ 580 \text{ kJ in 4 h}$$

1 kW is a heat extraction of 1 kJ/s. 4 h contain $4 \times 60 \times 60 = 14\ 400$ s.
Required heat extraction is therefore

$$\frac{2\ 580}{14\ 400} = 0\cdot179 \text{ kW}$$

Using the coldroom as in example A we get:

Heat penetration load = 0·480 kW
Service load (20%)　　= 0·096
Fan heat (75 W)　　　= 0·075
Product load　　　　 = 0·179
　　　　　　　　　　 0·830 kW

It is perhaps worth noting at this point that Imperial terms such as 'a 4 hp condensing unit', 'a 3 ton refrigeration load' or 'a heat extraction rate of 5 000 Btu/h' would all be expressed in kW in the SI system, thus avoiding a good deal of work converting from one energy unit to the other.

Example I

How much heat must be extracted from 50 kg of fresh lean beef to reduce its temperature from 18°C to −5°C?

From Table 19.4, using SI figures, it is seen that the SHC above freezing is 3·22 kJ/kg°C, latent heat is 232 kJ/kg°C and SHC below freezing 1·68 kJ/kg°C. Freezing point is −1·7°C.
Thus:

Sensible heat above freezing = $50 \times 3\cdot22 \times 19\cdot7$ = 　3 170
Latent heat　　　　　　　 = 50×232　　　　　 = 11 600
Sensible heat below freezing = $50 \times 1\cdot68 \times 3\cdot3$ = 　　277
　　　　　　　　　　　　　　　　　　　　　　　　　 15 047 kJ

Example J

0·5 t (500 kg) of veal at 7°C to be reduced to −6°C in not more than 12 h.

From SI table: SHCs are 2·97 and 1·63 kJ/kg°C
Latent heat is 212 kJ/kg
Freezing point is −1·7°C

Product load:

Sensible heat above freezing = 500 × 2·97 × 8·7 =   12 910
Latent heat   = 500 × 212   = 106 000
Sensible heat below freezing = 500 × 1·63 × 4·3 =   3 502
                                                   ─────────
                                                   122 412   kJ

12 h = 12 × 60 × 60 = 43 200 s.

Extraction rate for product load $= \dfrac{122\,412}{43\,200} = 2\cdot835$ kW

From example D:

Product load   = 2·835
Penetration load   = 0·404
10 % service load = 0·040
300 W fan load   = 0·300
                     ───────
                     3·579 kW

Example K

The total calculated load for a plant is found to be 101 000 kJ in 24 h. What capacity plant would be needed to give a running time of 16 h in 24 h?

16 h = 16 × 60 × 60 = 57 600 s

Required capacity $= \dfrac{101\,000}{57\,600} = 1\cdot75$ kW

STORAGE TEMPERATURES

TEST QUESTIONS

Test Questions No. 3, covering the third and last section of the book, are to be found on p. 143. The student is advised, as before, to check back on any point on which he is not absolutely clear.

TABLE 19.2 (*see* p. 125)

This table provides useful data concerning suitable storage temperatures for a selection of commodities, and it can be seen that

TABLE 19.1

Heat Conductivity of Building and Insulating Materials

The figures are given in k factors in Btu per square foot per hour for 1 in thickness and 1°F temperature difference, and kilo-calories per square metre per hour for 1 m thickness and 1°C temperature difference.

Material	Density		k		
	lb/ft³	kg/m³	Btu/ft²h°F	kcal/m²h°C	W/m°C Wm/m²°C
Brick wall—dry	110	1 762	8	0·99	1·15
Brick wall—wet	145	2 306	12	1·49	1·73
Concrete, ballast	65	1 025	10	1·24	1·44
Concrete, foamed slag	40	641	1·7	0·21	0·245
Concrete, cellular	60	961	1·0	0·12	0·144
			1·8	0·22	0·26
Plaster			4	0·50	0·58
Soil, loam (typical)	120	1 922	8 to 11	0·99 to 1·36	1·16 to 1·6
Soil, clay (typical)			10	1·24	1·44
Steel	480	7 688	300	37·2	43
Wood			0·8 to 1·1	0·99 to 0·13	0·12 to 0·16
Water			4·3	0·53	0·62
Ice			15	1·86	2·1
Cement, cellular	25	400	0·65	0·081	0·094
Cork, baked slab	7	112	0·26	0·032	0·037
	9	144	0·29	0·036	0·042
Cork, baked slab, wet	12	192	0·34	0·042	0·049
Cork, raw granulated	5 to 7	80 to 112	0·34	0·042	0·049
Cork, baked re-granulated	5 to 6	80 to 96	0·27	0·033	0·039
Cork, slab, bitumen binder	15	240	0·35	0·043	0·05
Cork, slab, cement binder	15	240	0·37	0·046	0·053

Ebonite, cellular	4 to 6	64 to 96	0·20	0·025	0·029
Glass wool, white	5	80	0·23	0·028	0·033
Glass wool, bitumen bonded	3 to 5	48 to 80	0·23	0·028	0·033
Kapok, cieba fibre	1	16	0·22	0·023	0·032
Mineral wool, felted	4	64	0·23	0·028	0·033
Polystyrene, cellular (small pores)	1·5	24	0·23	0·028	0·033
	2	32	0·21	0·026	0·030
	4	64	0·22	0·027	0·033
	5·5	88	0·24	0·030	0·035
Polyurethane (R. blown)	2 to 3		0·17		
Polyurethane, cellular	2·5		0·26		
PVC, cellular	3	48	0·25	0·031	0·036
	5	80	0·24	0·030	0·035
	7	112	0·27	0·033	0·039
	10	160	0·28	0·035	0·041
Slag wool, felted	8·5	136	0·23	0·028	0·033
Slag wool, loose	11	176	0·25	0·031	0·036
Urea formaldehyde, cellular	0·5	8	0·26	0·032	0·037
	1	16	0·22	0·023	0·032
	2	32	0·22	0·023	0·032
	3	48	0·23	0·028	0·033
	4	64	0·24	0·030	0·035
Vermiculite	9		0·47 to 0·58		
Aggregate	30		0·9		
	40		1·2		
	50		1·6		
Wallboard, insulating	20	320	0·4	0·050	0·060
Wood wool slabs	30	480	0·65	0·081	0·094

Based on data supplied by the National Physical Laboratory, Teddington.

TABLE 19.2

Cold Storage Temperatures

| Product | Temperature | |
	°C	°F
Meat		
Beef (chilled)	−2–0	29–32
Beef (frozen)	−18	0
Mutton (chilled)	−2–0	29–32
Mutton (frozen)	−18	0
Lamb (chilled)	−2–0	29–32
Lamb (frozen)	−18	0
Pork (chilled)	0–1	32–34
Pork (frozen)	−18	0
Bacon (chilled)	2–3	36–38
Bacon (frozen)	−18	0
Poultry (fresh)	−1–2	30–35
Poultry (frozen)	−21	−5
Game (fresh)	0–2	32–35
Game (frozen)	−21	−5
Rabbits (fresh)	0–2	30–35
Rabbits (frozen)	−18	0
Dairy Produce		
Butter (dairy)	0–2	32–35
Butter (frozen)	−10--−9	14–16
* Cheese	0–5	32–40
Milk	1	33
Margarine	3–5	38–40
Eggs (shell)	−1–1	31–33
Fish		
Fish (chilled)	−2--−1	28–30
Fish (frozen)	−21	−5 and below
Cereals		
Flour	6	42
Oatmeal	0–1	32–34
Fruit and Vegetables		
* Apples	1–5	34–40
Asparagus	2	35
Bananas	13	55
Berries (fresh)	0–1	32–34
Berries (frozen)	−18	0

TABLE 19.2—*contd*

Product	Temperature	
	°C	°F
Fruit and Vegetables		
Cabbage	0–2	32–35
Carrots	0–1	32–34
Celery	0–1	32–34
Cucumbers	5	40
Fruit (canned)	2	36
Fruit (dried)	5	40
Grapes	0–1	32–34
Lemons	10–13	50–55
Mushrooms	7–10	45–50
Onions	0–1	32–34
Oranges	3–5	38–40
Peaches	0–1	32–34
Pears	0	32
Plums	0–1	32–34
Potatoes	5–7	40–45
Tomatoes	10	50
Beverages		
Beer (barrels)	13–16	55–60
Beer (bottles)	1–9	33–48
Cider	0	32
Wines	10–13	50–55
Mineral Waters	2	36
Miscellaneous		
Bulbs	0–16	32–60
Cigars	3–7	37–45
* Flowers (cut)	2–5	35–40
Furs	−1–2	30–35
Honey	5–7	40–45
Hops	−2–1	28–34
Ice Cream	−18	0
Lard	5	40
Tobacco	5–7	40–50

Note: The storage temperature for frozen meat may vary with length of storage required.
* According to variety.

TABLE 19.3
Specific Heat, Btu/lb°F

Product	Humidity % RH	Above freezing	Below freezing	Latent heat	Freezing point	Respiration Btu/lb day
Apples	85–88	0·86	0·45	121	28·4	0·72
Asparagus	85–90	0·94	0·48	134	29·8	5·75
Bacon, fresh	80	0·50	0·30	29	25	
Bananas	85–95	0·80	0·42	108	28	4·18
Beans, green	85–90	0·91	0·47	128	29·7	3·37
Beans, dried	70	0·30	0·24	18	28	
Beef, fresh, fat	84	0·60	0·35	79	28	
Beef, fresh, lean	85	0·77	0·40	100	29	
Beets	95–98	0·86	0·47	129	31·1	0·91
Blackberries	80–85	0·88	0·46	122	28·9	
Broccoli	90–95	0·92	0·47	130	29·2	
Butter		0·64	0·34	15	30	
Cabbage	90–95	0·94	0·47	132	31·2	
Carrots	95–98	0·86	0·45	126	29·6	1·73
Cauliflower	85–90	0·93	0·47	132	30·1	
Celery	90–95	0·95	0·48	135	29·7	2·27
Cheese, American		0·64	0·36	79	17	2·34
Cherries	80–85	0·87	0·45	120	26	1·32
Cream (40%)		0·85	0·40	90	28	
Cucumbers	80–85	0·97	0·49	137	30·5	
Eggs		0·76	0·40	100	27	
Fish, fresh, iced		0·76	0·41	101	30	
Fish, dried	60–70	0·56	0·34	65		
Furs	40–60		0·40			
Grapes	80–85	0·88	0·44	116	26·3	0·42
Ham, fresh	80	0·68	0·38	86·7	27	
Ice Cream		0·78	0·45	96	27	
Lard	80	0·52	0·31	90		
Lamb	82	0·67	0·30	83·5	29	
Lettuce	90–95	0·96	0·48	136	31·1	3·69
Liver, fresh	83	0·72	0·40	93·3	29	
Meat, brined		0·75	0·36	75		
Milk		0·93	0·49	124	31	
Mushrooms	80–85	0·93	0·47	130	30·2	4·0
Peas, green	85–90	0·79	0·42	106	30	
Pork, fresh	85	0·68	0·38	86·5	28	
Poultry		0·79	0·37	106	27	
Sausage, fresh	80	0·89	0·56	93	26	
Strawberries	80–85	0·93	0·47	129	29·9	3·3
Veal		0·71	0·39	91	29	
Vegetables, mixed	90–95	0·90	0·45	130	30	2

* Details extracted from: Table No. 6 Engineering Data, by kind permission of Searle Manufacturing Co. Ltd, Newgate Lane, Fareham, Hants.

TABLE 19.4

Specific Heat Capacity, kJ/kg°C

Product	Humidity % RH	Above freezing	Below freezing	Latent heat	Freezing point	Respiration kJ/kg day
Apples	85–88	3·60	1·88	281	−2·0	1·67
Asparagus	85–90	3·93	2·01	312	−1·2	13·36
Bacon, fresh	80	2·09	1·26	67	−3·9	
Bananas	85–95	3·35	1·76	251	−2·2	9·70
Beans, green	85–90	3·80	1·97	298	−1·3	7·84
Beans, dried	70	1·25	1·00	42	−2·2	
Beef, fresh, fat	84	2·57	1·46	184	−2·2	
Beef, fresh, lean	85	3·22	1·68	232	−1·7	
Beets	95–98	3·60	1·97	300	−0·5	2·12
Blackberries	80–85	3·68	1·92	284	−1·7	
Broccoli	90–95	3·85	1·97	302	−1·5	
Butter		2·68	1·42	35	−1·1	
Cabbage	90–95	3·93	1·97	308	−0·5	
Carrots	95–98	3·60	1·88	293	−1·3	4·02
Cauliflower	85–90	3·89	1·97	307	−1·0	
Celery	90–95	3·97	2·01	314	−1·3	5·28
Cheese, American		2·68	1·51	184	−8·3	5·44
Cherries	80–85	3·64	1·88	279	−3·3	3·07
Cream (40%)		3·56	1·68	209	−2·2	
Cucumbers	80–85	4·05	2·05	318	−0·8	
Eggs		3·18	1·68	232	−2·8	
Fish, fresh, iced		3·18	1·72	234	−1·1	
Fish, dried	60–70	2·34	1·42	151		
Furs	40–60		1·68			
Grapes	80–85	3·68	1·84	270	−3·2	0·98
Ham, fresh	80	2·84	1·59	202	−2·8	
Ice Cream		3·26	1·88	223	−2·8	
Lard	80	2·18	1·30	209		
Lamb	82	2·80	1·25	194	−1·7	
Lettuce	90–95	4·01	2·01	316	−0·5	8·56
Liver, fresh	83	3·01	1·68	217	−1·7	
Meat, brined		3·14	1·51	174		
Milk		3·88	2·05	288	−0·6	
Mushrooms	80–85	3·88	1·97	302	−1·0	9·3
Peas, green	85–90	3·30	1·76	246	−1·1	
Pork, fresh	85	2·84	1·59	201	−2·2	
Poultry		3·30	1·55	246	−2·8	
Sausage, fresh	80	3·72	2·34	216	−3·3	
Strawberries	80–85	3·89	1·97	300	−1·2	7·7
Veal		2·97	1·63	212	−1·7	
Vegetables, mixed	90–95	3·77	1·88	302	−1·1	5

different products require a wide range of different storage temperatures. For example, beer is stored at 13°–16°C (55°–60°F), poultry 1°–2°C (30°–35°F), and ice cream −18°C (0°F).

From this may be obtained the information required in designing a coldroom to suit a customer's requirements.

20
Servicing Notes

This last chapter is concerned with the more general essentials of a refrigeration engineer, which although not technical, are none the less important. They are too often neglected or even taken for granted, and if some of these aspects have been dealt with already, they are of sufficient consequence to be repeated.

For obvious reasons an apprentice will not at first be expected to do service work alone, but sometimes the engineer he accompanies, whilst being well equipped technically, has little concept of the finer points of servicing, and is not therefore an ideal tutor in this respect. It is hoped the following observations will assist in some small way to give an apprentice a few thoughts on the matter, so that he may develop his own individual approach, and not be misguided by poor example. So for the moment the student is asked to look ahead to the time when he will be allocated a van, and venture into the world of refrigeration service on his own.

The first thing is to make an early start to the day, that is to say have all the service sheets from the day before made up ready to present to the service manager, plus any other regulation regarding renewal of parts, time sheets, etc., prepared so that a minimum of time is spent on these details. The instruction made earlier in the book, to make up the work sheets on site, is based on the common-sense notion that facts are then fresh in the mind, that details of spare parts used are less likely to be forgotten, and that there will be no need to make them up hurriedly *en bloc* in the morning. The service manager will also appreciate his attention being drawn to any feature of a job which may require action being taken, or to elaborate on the information contained on the sheet.

Having set off, with spares renewed, and a new batch of service sheets, what does a good service engineer take with him? First a respectable appearance, smart overalls and a regularly washed van. His tool-box should be clean, tidy and well stocked; gauges protected and kept separate from other tools; and any special instruments he may be carrying, such as electronic leak detector, megger, etc., must be placed in such a way as to avoid being damaged. Expensive appliances must always be treated with extra care.

His stock of spares must always be kept up to date, he must be familiar with all the parts allocated to the vehicle and know where to find them. Gas cylinders must always be adequately filled ready for the day ahead, and other accessories such as oil regularly checked. Service sheets should be neatly secured to a board with a bull-dog clip, and not stuffed into an inside pocket, or loosely spread about the van. Any other literature, such as instruction manuals, must be stored where they are accessible and readable when required for reference.

An engineer also takes with him his skill and knowledge, and this he must add to at every opportunity by reading and studying special information, by discussing with his colleagues problems met in the field, and by learning from his own experience. Learning never stops. New equipment, new applications, new methods are constantly being introduced so that knowledge at no time remains static, and every engineer must be ready to try new ideas in an effort to improve the service he gives.

The service engineer is entrusted with a vehicle, and he is responsible for looking after it. Tyres and batteries must be checked at least once a week, and instructions on regular servicing observed. It is his duty to treat the van with respect and drive carefully. Van breakdowns, which always seem to occur at the most inopportune moment, are too often the result of reckless driving. Brakes, clutch and tyres wear out more from abuse than use, and expenses incurred in this way are an unnecessary overhead for the company. By far the most wear on a vehicle takes place on starting, stopping, gear-changing and cornering, and an engineer of all people, should, in consideration of machinery, carry out these operations gently. A driver of a vehicle is also responsible for observing the laws of the road, including parking restrictions and so on.

Having reached his destination, the engineer should enter the premises briskly with his tool-box, and not amble in as if he just happened to be passing. He must then report to the person in charge, not only to obtain as much information as possible, but to ensure that his presence is known, and then to carry out a brief but thorough observation of the plant. From this preliminary check, and consideration of other facts, will follow the further procedure deemed necessary. If this involves spare parts, inform the customer before fitting them. Consultation at all stages with the customer avoids remonstrations from him when presented with the account.

During the repair a sense of urgency, and an awareness of time, must always be maintained. This does not imply that the job must be rushed or skimped, but simply that the invoice to the client will be related to the time taken and this must be justified by the work done.

Whilst waiting for response to adjustment or fitting of new parts, an opportunity is given to repair minor defects such as pipe rattle, vibrating drier, loose clips and so on. The condition of electric leads should be noted and attended to if considered necessary. It is during this period too that the work sheet can be written up, giving concise details of the fault and action taken, together with spare parts used. Any other useful information should also be recorded.

Before leaving, the engineer should always make a last inspection to make certain everything is in order, that terminal covers for example have been replaced, tools collected, and that any debris resulting from the repair has been cleaned up. Finally he must have a last word with the manager and *obtain his signature*.

During the course of the day it is obviously necessary for the engineer to maintain regular contact with his headquarters, both to inform the service manager of his progress, and to receive any instructions regarding a change of programme. Normally some system of telephoning is laid down to meet this necessity.

Refrigeration is a young industry, growing daily more vital to our way of life. Blood banks, computers, industrial processes, air conditioning and many other fields rely very much on mechanical cooling. The refrigeration engineer is therefore providing an indispensable service to the community, pursuing a trade of which he has every right to be proud. To sustain the service its status deserves, refrigeration needs a constant flow of alert, skilled men, dedicated to maintaining refrigeration at all times. This book has endeavoured to give guidance to the training of such men.

Glossary

ABSOLUTE PRESSURE
Pressure starting from a maximum vacuum. Gauge pressure + 1 bar (15 lb).

AMBIENT
Surrounding temperature.

ATMOSPHERIC PRESSURE
The weight of air pressing on the earth. 1 bar (15 lb)/in^2 at sea level.

BACK PRESSURE
The pressure registering on the suction gauge of a compressor.

BACK SEATED
A compressor service valve turned fully in an anti-clockwise direction.

BOILING POINT
The temperature at which a liquid is transferred into a gas.

Btu
The quantity of heat required to be given to or taken from 1 lb of water in raising or lowering its temperature by 1°F.

CALORIE
The quantity of heat required to be given to or taken from 1 g of water in raising or lowering its temperature by 1°C.

FLARE
The spreading of a tube end to form a join.

FLARE BLOCK
A tool designed to make a flare.

FREEZING POINT
The temperature at which a liquid is transferred into a solid.

FRONT SEATED
A compressor service valve turned fully in a clockwise direction.

HEAD PRESSURE
The pressure registering on the discharge gauge of a compressor.

k FACTOR—IMPERIAL UNITS
The insulating value of a material, represented by the quantity of heat in Btu/h which will pass through 1 ft^2 of the material, 1 in thick, when subjected to a temperature difference of 1°F.

k FACTOR—SI UNITS
The quantity of heat measured in watts which will pass through a 1 m cube of the material when subjected to a temperature difference of 1°C.

LATENT HEAT OF EVAPORATION
The quantity of heat required to be given to or taken from a substance to change its state from a liquid to a gas, or gas to a liquid, without changing its temperature.

LATENT HEAT OF FUSION
The quantity of heat required to be given to or taken from a substance to change its state from liquid to solid, or solid to liquid, without changing its temperature.

LAW OF THERMODYNAMICS (FIRST)
Heat and work are mutually convertible, and each may be expressed in terms of the other.

LAW OF THERMODYNAMICS (SECOND)
Heat will always travel from a hot body to a cooler one.

MEGGER
An instrument for testing the insulation resistance of an electrical circuit.

PUMPING DOWN
The action of shutting the liquid receiver outlet valve, so that the refrigerant cannot pass beyond it. Refrigerant in the system is thus liquefied in the condenser and retained in the liquid receiver.

PURGING

The process of allowing refrigerant gas to pass through piping or other parts of a system to expel air.

RH (RELATIVE HUMIDITY)

The ratio of aqueous vapour in the atmosphere, at any given temperature, to that which it could hold at that temperature, if saturated.

SATURATED VAPOUR

Vapour containing its maximum quantity of liquid particles in suspension.

SENSIBLE HEAT

Heat which can be seen to be given to a substance by observing its rise in temperature.

SPECIFIC HEAT

The ratio of heat required to change the temperature of a substance, to the amount required to change an equal mass of water the same number of degrees.

SPECIFIC HEAT CAPACITY

The quantity of heat measured in kilojoules required to raise the temperature of 1 kg of a substance through 1°C.

TD (TEMPERATURE DIFFERENCE)

Used in refrigeration to indicate the difference in temperature between air or gas, and the medium by which it is being cooled.

VACUUM

A pressure less than that of the prevailing atmospheric pressure.

Logarithms

The following Tables are reproduced from Frank Castle's *Logarithmic and other Tables* by kind permission of Macmillan and Co. Ltd.

LOGARITHMS

	0	1	2	3	4	5	6	7	8	9	1 2 3	4 5 6	7 8 9
10	0000	0043	0086	0128	0170	0212	0253	0294	0334	0374	5 9 13 / 4 8 12	17 21 26 / 16 20 24	30 34 38 / 28 32 36
11	0414	0453	0492	0531	0569	0607	0645	0682	0719	0755	4 8 12 / 4 7 11	16 20 23 / 15 18 22	27 31 35 / 26 29 33
12	0792	0828	0864	0899	0934	0969	1004	1038	1072	1106	3 7 11 / 3 7 10	14 18 21 / 14 17 20	25 28 32 / 24 27 31
13	1139	1173	1206	1239	1271	1303	1335	1367	1399	1430	3 6 10 / 3 7 10	13 16 19 / 13 16 19	23 26 29 / 22 26 29
14	1461	1492	1523	1553	1584	1614	1644	1673	1703	1732	3 6 9 / 3 6 9	12 15 19 / 12 14 17	22 25 28 / 20 23 26
15	1761	1790	1818	1847	1875	1903	1931	1959	1987	2014	3 6 9 / 3 6 8	11 14 17 / 11 14 17	20 23 26 / 19 22 25
16	2041	2068	2095	2122	2148	2175	2201	2227	2253	2279	3 6 8 / 3 5 8	11 14 16 / 10 13 16	19 22 24 / 18 21 23
17	2304	2330	2355	2380	2405	2430	2455	2480	2504	2529	3 5 8 / 3 5 8	10 13 15 / 10 12 15	18 20 23 / 17 20 22
18	2553	2577	2601	2625	2648	2672	2695	2718	2742	2765	2 5 7 / 2 4 7	9 12 14 / 9 11 14	16 18 21 / 16 18 21
19	2788	2810	2833	2856	2878	2900	2923	2945	2967	2989	2 4 7 / 2 4 6	9 11 13 / 8 11 13	16 18 20 / 15 17 19
20	3010	3032	3054	3075	3096	3118	3139	3160	3181	3201	2 4 6	8 11 13	15 17 19
21	3222	3243	3263	3284	3304	3324	3345	3365	3385	3404	2 4 6	8 10 12	14 16 18
22	3424	3444	3464	3483	3502	3522	3541	3560	3579	3598	2 4 6	8 10 12	14 15 17
23	3617	3636	3655	3674	3692	3711	3729	3747	3766	3784	2 4 6	7 9 11	13 15 17
24	3802	3820	3838	3856	3874	3892	3909	3927	3945	3962	2 4 5	7 9 11	12 14 16
25	3979	3997	4014	4031	4048	4065	4082	4099	4116	4133	2 3 5	7 9 10	12 14 15
26	4150	4166	4183	4200	4216	4232	4249	4265	4281	4298	2 3 5	7 8 10	11 13 15
27	4314	4330	4346	4362	4378	4393	4409	4425	4440	4456	2 3 5	6 8 9	11 13 14
28	4472	4487	4502	4518	4533	4548	4564	4579	4594	4609	2 3 5	6 8 9	11 12 14
29	4624	4639	4654	4669	4683	4698	4713	4728	4742	4757	1 3 4	6 7 9	10 12 13
30	4771	4786	4800	4814	4829	4843	4857	4871	4886	4900	1 3 4	6 7 9	10 11 13
31	4914	4928	4942	4955	4969	4983	4997	5011	5024	5038	1 3 4	6 7 8	10 11 12
32	5051	5065	5079	5092	5105	5119	5132	5145	5159	5172	1 3 4	5 7 8	9 11 12
33	5185	5198	5211	5224	5237	5250	5263	5276	5289	5302	1 3 4	5 6 8	9 10 12
34	5315	5328	5340	5353	5366	5378	5391	5403	5416	5428	1 3 4	5 6 8	9 10 11
35	5441	5453	5465	5478	5490	5502	5514	5527	5539	5551	1 2 4	5 6 7	9 10 11
36	5563	5575	5587	5599	5611	5623	5635	5647	5658	5670	1 2 4	5 6 7	8 10 11
37	5682	5694	5705	5717	5729	5740	5752	5763	5775	5786	1 2 3	5 6 7	8 9 10
38	5798	5809	5821	5832	5843	5855	5866	5877	5888	5899	1 2 3	5 6 7	8 9 10
39	5911	5922	5933	5944	5955	5966	5977	5988	5999	6010	1 2 3	4 5 7	8 9 10
40	6021	6031	6042	6053	6064	6075	6085	6096	6107	6117	1 2 3	4 5 6	8 9 10
41	6128	6138	6149	6160	6170	6180	6191	6201	6212	6222	1 2 3	4 5 6	7 8 9
42	6232	6243	6253	6263	6274	6284	6294	6304	6314	6325	1 2 3	4 5 6	7 8 9
43	6335	6345	6355	6365	6375	6385	6395	6405	6415	6425	1 2 3	4 5 6	7 8 9
44	6435	6444	6454	6464	6474	6484	6493	6503	6513	6522	1 2 3	4 5 6	7 8 9
45	6532	6542	6551	6561	6571	6580	6590	6599	6609	6618	1 2 3	4 5 6	7 8 9
46	6628	6637	6646	6656	6665	6675	6684	6693	6702	6712	1 2 3	4 5 6	7 7 8
47	6721	6730	6739	6749	6758	6767	6776	6785	6794	6803	1 2 3	4 5 5	6 7 8
48	6812	6821	6830	6839	6848	6857	6866	6875	6884	6893	1 2 3	4 4 5	6 7 8
49	6902	6911	6920	6928	6937	6946	6955	6964	6972	6981	1 2 3	4 4 5	6 7 8

LOGARITHMS

	0	1	2	3	4	5	6	7	8	9	123	456	789
50	6990	6998	7007	7016	7024	7033	7042	7050	7059	7067	1 2 3	3 4 5	6 7 8
51	7076	7084	7093	7101	7110	7118	7126	7135	7143	7152	1 2 3	3 4 5	6 7 8
52	7160	7168	7177	7185	7193	7202	7210	7218	7226	7235	1 2 2	3 4 5	6 7 7
53	7243	7251	7259	7267	7275	7284	7292	7300	7308	7316	1 2 2	3 4 5	6 6 7
54	7324	7332	7340	7348	7356	7364	7372	7380	7388	7396	1 2 2	3 4 5	6 6 7
55	7404	7412	7419	7427	7435	7443	7451	7459	7466	7474	1 2 2	3 4 5	5 6 7
56	7482	7490	7497	7505	7513	7520	7528	7536	7543	7551	1 2 2	3 4 5	5 6 7
57	7559	7566	7574	7582	7589	7597	7604	7612	7619	7627	1 2 2	3 4 5	5 6 7
58	7634	7642	7649	7657	7664	7672	7679	7686	7694	7701	1 1 2	3 4 4	5 6 7
59	7709	7716	7723	7731	7738	7745	7752	7760	7767	7774	1 1 2	3 4 4	5 6 7
60	7782	7789	7796	7803	7810	7818	7825	7832	7839	7846	1 1 2	3 4 4	5 6 6
61	7853	7860	7868	7875	7882	7889	7896	7903	7910	7917	1 1 2	3 4 4	5 6 6
62	7924	7931	7938	7945	7952	7959	7966	7973	7980	7987	1 1 2	3 3 4	5 6 6
63	7993	8000	8007	8014	8021	8028	8035	8041	8048	8055	1 1 2	3 3 4	5 5 6
64	8062	8069	8075	8082	8089	8096	8102	8109	8116	8122	1 1 2	3 3 4	5 5 6
65	8129	8136	8142	8149	8156	8162	8169	8176	8182	8189	1 1 2	3 3 4	5 5 6
66	8195	8202	8209	8215	8222	8228	8235	8241	8248	8254	1 1 2	3 3 4	5 5 6
67	8261	8267	8274	8280	8287	8293	8299	8306	8312	8319	1 1 2	3 3 4	5 5 6
68	8325	8331	8338	8344	8351	8357	8363	8370	8376	8382	1 1 2	3 3 4	4 5 6
69	8388	8395	8401	8407	8414	8420	8426	8432	8439	8445	1 1 2	2 3 4	4 5 6
70	8451	8457	8463	8470	8476	8482	8488	8494	8500	8506	1 1 2	2 3 4	4 5 6
71	8513	8519	8525	8531	8537	8543	8549	8555	8561	8567	1 1 2	2 3 4	4 5 5
72	8573	8579	8585	8591	8597	8603	8609	8615	8621	8627	1 1 2	2 3 4	4 5 5
73	8633	8639	8645	8651	8657	8663	8669	8675	8681	8686	1 1 2	2 3 4	4 5 5
74	8692	8698	8704	8710	8716	8722	8727	8733	8739	8745	1 1 2	2 3 4	4 5 5
75	8751	8756	8762	8768	8774	8779	8785	8791	8797	8802	1 1 2	2 3 3	4 5 5
76	8808	8814	8820	8825	8831	8837	8842	8848	8854	8859	1 1 2	2 3 3	4 5 5
77	8865	8871	8876	8882	8887	8893	8899	8904	8910	8915	1 1 2	2 3 3	4 4 5
78	8921	8927	8932	8938	8943	8949	8954	8960	8965	8971	1 1 2	2 3 3	4 4 5
79	8976	8982	8987	8993	8998	9004	9009	9015	9020	9025	1 1 2	2 3 3	4 4 5
80	9031	9036	9042	9047	9053	9058	9063	9069	9074	9079	1 1 2	2 3 3	4 4 5
81	9085	9090	9096	9101	9106	9112	9117	9122	9128	9133	1 1 2	2 3 3	4 4 5
82	9138	9143	9149	9154	9159	9165	9170	9175	9180	9186	1 1 2	2 3 3	4 4 5
83	9191	9196	9201	9206	9212	9217	9222	9227	9232	9238	1 1 2	2 3 3	4 4 5
84	9243	9248	9253	9258	9263	9269	9274	9279	9284	9289	1 1 2	2 3 3	4 4 5
85	9294	9299	9304	9309	9315	9320	9325	9330	9335	9340	1 1 2	2 3 3	4 4 5
86	9345	9350	9355	9360	9365	9370	9375	9380	9385	9390	1 1 2	2 3 3	4 4 5
87	9395	9400	9405	9410	9415	9420	9425	9430	9435	9440	0 1 1	2 2 3	3 4 4
88	9445	9450	9455	9460	9465	9469	9474	9479	9484	9489	0 1 1	2 2 3	3 4 4
89	9494	9499	9504	9509	9513	9518	9523	9528	9533	9538	0 1 1	2 2 3	3 4 4
90	9542	9547	9552	9557	9562	9566	9571	9576	9581	9586	0 1 1	2 2 3	3 4 4
91	9590	9595	9600	9605	9609	9614	9619	9624	9628	9633	0 1 1	2 2 3	3 4 4
92	9638	9643	9647	9652	9657	9661	9666	9671	9675	9680	0 1 1	2 2 3	3 4 4
93	9685	9689	9694	9699	9703	9708	9713	9717	9722	9727	0 1 1	2 2 3	3 4 4
94	9731	9736	9741	9745	9750	9754	9759	9763	9768	9773	0 1 1	2 2 3	3 4 4
95	9777	9782	9786	9791	9795	9800	9805	9809	9814	9818	0 1 1	2 2 3	3 4 4
96	9823	9827	9832	9836	9841	9845	9850	9854	9859	9863	0 1 1	2 2 3	3 4 4
97	9868	9872	9877	9881	9886	9890	9894	9899	9903	9908	0 1 1	2 2 3	3 4 4
98	9912	9917	9921	9926	9930	9934	9939	9943	9948	9952	0 1 1	2 2 3	3 4 4
99	9956	9961	9965	9969	9974	9978	9983	9987	9991	9996	0 1 1	2 2 3	3 3 4

ANTILOGARITHMS

	0	1	2	3	4	5	6	7	8	9	123	456	789
·00	1000	1002	1005	1007	1009	1012	1014	1016	1019	1021	0 0 1	1 1 1	2 2 2
·01	1023	1026	1028	1030	1033	1035	1038	1040	1042	1045	0 0 1	1 1 1	2 2 2
·02	1047	1050	1052	1054	1057	1059	1062	1064	1067	1069	0 0 1	1 1 1	2 2 2
·03	1072	1074	1076	1079	1081	1084	1086	1089	1091	1094	0 0 1	1 1 1	2 2 2
·04	1096	1099	1102	1104	1107	1109	1112	1114	1117	1119	0 1 1	1 1 2	2 2 2
·05	1122	1125	1127	1130	1132	1135	1138	1140	1143	1146	0 1 1	1 1 2	2 2 2
·06	1148	1151	1153	1156	1159	1161	1164	1167	1169	1172	0 1 1	1 1 2	2 2 2
·07	1175	1178	1180	1183	1186	1189	1191	1194	1197	1199	0 1 1	1 1 2	2 2 2
·08	1202	1205	1208	1211	1213	1216	1219	1222	1225	1227	0 1 1	1 1 2	2 2 3
·09	1230	1233	1236	1239	1242	1245	1247	1250	1253	1256	0 1 1	1 1 2	2 2 3
·10	1259	1262	1265	1268	1271	1274	1276	1279	1282	1285	0 1 1	1 1 2	2 2 3
·11	1288	1291	1294	1297	1300	1303	1306	1309	1312	1315	0 1 1	1 2 2	2 2 3
·12	1318	1321	1324	1327	1330	1334	1337	1340	1343	1346	0 1 1	1 2 2	2 2 3
·13	1349	1352	1355	1358	1361	1365	1368	1371	1374	1377	0 1 1	1 2 2	2 3 3
·14	1380	1384	1387	1390	1393	1396	1400	1403	1406	1409	0 1 1	1 2 2	2 3 3
·15	1413	1416	1419	1422	1426	1429	1432	1435	1439	1442	0 1 1	1 2 2	2 3 3
·16	1445	1449	1452	1455	1459	1462	1466	1469	1472	1476	0 1 1	1 2 2	2 3 3
·17	1479	1483	1486	1489	1493	1496	1500	1503	1507	1510	0 1 1	1 2 2	2 3 3
·18	1514	1517	1521	1524	1528	1531	1535	1538	1542	1545	0 1 1	1 2 2	2 3 3
·19	1549	1552	1556	1560	1563	1567	1570	1574	1578	1581	0 1 1	1 2 2	3 3 3
·20	1585	1589	1592	1596	1600	1603	1607	1611	1614	1618	0 1 1	1 2 2	3 3 3
·21	1622	1626	1629	1633	1637	1641	1644	1648	1652	1656	0 1 1	2 2 2	3 3 3
·22	1660	1663	1667	1671	1675	1679	1683	1687	1690	1694	0 1 1	2 2 2	3 3 3
·23	1698	1702	1706	1710	1714	1718	1722	1726	1730	1734	0 1 1	2 2 2	3 3 4
·24	1738	1742	1746	1750	1754	1758	1762	1766	1770	1774	0 1 1	2 2 2	3 3 4
·25	1778	1782	1786	1791	1795	1799	1803	1807	1811	1816	0 1 1	2 2 2	3 3 4
·26	1820	1824	1828	1832	1837	1841	1845	1849	1854	1858	0 1 1	2 2 3	3 3 4
·27	1862	1866	1871	1875	1879	1884	1888	1892	1897	1901	0 1 1	2 2 3	3 3 4
·28	1905	1910	1914	1919	1923	1928	1932	1936	1941	1945	0 1 1	2 2 3	3 4 4
·29	1950	1954	1959	1963	1968	1972	1977	1982	1986	1991	0 1 1	2 2 3	3 4 4
·30	1995	2000	2004	2009	2014	2018	2023	2028	2032	2037	0 1 1	2 2 3	3 4 4
·31	2042	2046	2051	2056	2061	2065	2070	2075	2080	2084	0 1 1	2 2 3	3 4 4
·32	2089	2094	2099	2104	2109	2113	2118	2123	2128	2133	0 1 1	2 2 3	3 4 4
·33	2138	2143	2148	2153	2158	2163	2168	2173	2178	2183	0 1 1	2 2 3	3 4 4
·34	2188	2193	2198	2203	2208	2213	2218	2223	2228	2234	1 1 2	2 3 3	4 4 5
·35	2239	2244	2249	2254	2259	2265	2270	2275	2280	2286	1 1 2	2 3 3	4 4 5
·36	2291	2296	2301	2307	2312	2317	2323	2328	2333	2339	1 1 2	2 3 3	4 4 5
·37	2344	2350	2355	2360	2366	2371	2377	2382	2388	2393	1 1 2	2 3 3	4 4 5
·38	2399	2404	2410	2415	2421	2427	2432	2438	2443	2449	1 1 2	2 3 3	4 4 5
·39	2455	2460	2466	2472	2477	2483	2489	2495	2500	2506	1 1 2	2 3 3	4 5 5
·40	2512	2518	2523	2529	2535	2541	2547	2553	2559	2564	1 1 2	2 3 4	4 5 5
·41	2570	2576	2582	2588	2594	2600	2606	2612	2618	2624	1 1 2	2 3 4	4 5 5
·42	2630	2636	2642	2649	2655	2661	2667	2673	2679	2685	1 1 2	2 3 4	4 5 6
·43	2692	2698	2704	2710	2716	2723	2729	2735	2742	2748	1 1 2	3 3 4	4 5 6
·44	2754	2761	2767	2773	2780	2786	2793	2799	2805	2812	1 1 2	3 3 4	4 5 6
·45	2818	2825	2831	2838	2844	2851	2858	2864	2871	2877	1 1 2	3 3 4	5 5 6
·46	2884	2891	2897	2904	2911	2917	2924	2931	2938	2944	1 1 2	3 3 4	5 5 6
·47	2951	2958	2965	2972	2979	2985	2992	2999	3006	3013	1 1 2	3 3 4	5 5 6
·48	3020	3027	3034	3041	3048	3055	3062	3069	3076	3083	1 1 2	3 4 4	5 6 6
·49	3090	3097	3105	3112	3119	3126	3133	3141	3148	3155	1 1 2	3 4 4	5 6 6

ANTILOGARITHMS

	0	1	2	3	4	5	6	7	8	9	1 2 3	4 5 6	7 8 9
·50	3162	3170	3177	3184	3192	3199	3206	3214	3221	3228	1 1 2	3 4 4	5 6 7
·51	3236	3243	3251	3258	3266	3273	3281	3289	3296	3304	1 2 2	3 4 5	5 6 7
·52	3311	3319	3327	3334	3342	3350	3357	3365	3373	3381	1 2 2	3 4 5	5 6 7
·53	3388	3396	3404	3412	3420	3428	3436	3443	3451	3459	1 2 2	3 4 5	6 6 7
·54	3467	3475	3483	3491	3499	3508	3516	3524	3532	3540	1 2 2	3 4 5	6 6 7
·55	3548	3556	3565	3573	3581	3589	3597	3606	3614	3622	1 2 2	3 4 5	6 7 7
·56	3631	3639	3648	3656	3664	3673	3681	3690	3698	3707	1 2 3	3 4 5	6 7 8
·57	3715	3724	3733	3741	3750	3758	3767	3776	3784	3793	1 2 3	3 4 5	6 7 8
·58	3802	3811	3819	3828	3837	3846	3855	3864	3873	3882	1 2 3	4 4 5	6 7 8
·59	3890	3899	3908	3917	3926	3936	3945	3954	3963	3972	1 2 3	4 5 5	6 7 8
·60	3981	3990	3999	4009	4018	4027	4036	4046	4055	4064	1 2 3	4 5 6	6 7 8
·61	4074	4083	4093	4102	4111	4121	4130	4140	4150	4159	1 2 3	4 5 6	7 8 9
·62	4169	4178	4188	4198	4207	4217	4227	4236	4246	4256	1 2 3	4 5 6	7 8 9
·63	4266	4276	4285	4295	4305	4315	4325	4335	4345	4355	1 2 3	4 5 6	7 8 9
·64	4365	4375	4385	4395	4406	4416	4426	4436	4446	4457	1 2 3	4 5 6	7 8 9
·65	4467	4477	4487	4498	4508	4519	4529	4539	4550	4560	1 2 3	4 5 6	7 8 9
·66	4571	4581	4592	4603	4613	4624	4634	4645	4656	4667	1 2 3	4 5 6	7 9 10
·67	4677	4688	4699	4710	4721	4732	4742	4753	4764	4775	1 2 3	4 5 7	8 9 10
·68	4786	4797	4808	4819	4831	4842	4853	4864	4875	4887	1 2 3	4 6 7	8 9 10
·69	4898	4909	4920	4932	4943	4955	4966	4977	4989	5000	1 2 3	5 6 7	8 9 10
·70	5012	5023	5035	5047	5058	5070	5082	5093	5105	5117	1 2 4	5 6 7	8 9 11
·71	5129	5140	5152	5164	5176	5188	5200	5212	5224	5236	1 2 4	5 6 7	8 10 11
·72	5248	5260	5272	5284	5297	5309	5321	5333	5346	5358	1 2 4	5 6 7	9 10 11
·73	5370	5383	5395	5408	5420	5433	5445	5458	5470	5483	1 3 4	5 6 8	9 10 11
·74	5495	5508	5521	5534	5546	5559	5572	5585	5598	5610	1 3 4	5 6 8	9 10 12
·75	5623	5636	5649	5662	5675	5689	5702	5715	5728	5741	1 3 4	5 7 8	9 10 12
·76	5754	5768	5781	5794	5808	5821	5834	5848	5861	5875	1 3 4	5 7 8	9 11 12
·77	5888	5902	5916	5929	5943	5957	5970	5984	5998	6012	1 3 4	5 7 8	10 11 12
·78	6026	6039	6053	6067	6081	6095	6109	6124	6138	6152	1 3 4	6 7 8	10 11 13
·79	6166	6180	6194	6209	6223	6237	6252	6266	6281	6295	1 3 4	6 7 9	10 11 13
·80	6310	6324	6339	6353	6368	6383	6397	6412	6427	6442	1 3 4	6 7 9	10 12 13
·81	6457	6471	6486	6501	6516	6531	6546	6561	6577	6592	2 3 5	6 8 9	11 12 14
·82	6607	6622	6637	6653	6668	6683	6699	6714	6730	6745	2 3 5	6 8 9	11 12 14
·83	6761	6776	6792	6808	6823	6839	6855	6871	6887	6902	2 3 5	6 8 9	11 13 14
·84	6918	6934	6950	6966	6982	6998	7015	7031	7047	7063	2 3 5	6 8 10	11 13 15
·85	7079	7096	7112	7129	7145	7161	7178	7194	7211	7228	2 3 5	7 8 10	12 13 15
·86	7244	7261	7278	7295	7311	7328	7345	7362	7379	7396	2 3 5	7 8 10	12 13 15
·87	7413	7430	7447	7464	7482	7499	7516	7534	7551	7568	2 3 5	7 9 10	12 14 16
·88	7586	7603	7621	7638	7656	7674	7691	7709	7727	7745	2 4 5	7 9 11	12 14 16
·89	7762	7780	7798	7816	7834	7852	7870	7889	7907	7925	2 4 5	7 9 11	13 14 16
·90	7943	7962	7980	7998	8017	8035	8054	8072	8091	8110	2 4 6	7 9 11	13 15 17
·91	8128	8147	8166	8185	8204	8222	8241	8260	8279	8299	2 4 6	8 9 11	13 15 17
·92	8318	8337	8356	8375	8395	8414	8433	8453	8472	8492	2 4 6	8 10 12	14 15 17
·93	8511	8531	8551	8570	8590	8610	8630	8650	8670	8690	2 4 6	8 10 12	14 16 18
·94	8710	8730	8750	8770	8790	8810	8831	8851	8872	8892	2 4 6	8 10 12	14 16 18
·95	8913	8933	8954	8974	8995	9016	9036	9057	9078	9099	2 4 6	8 10 12	15 17 19
·96	9120	9141	9162	9183	9204	9226	9247	9268	9290	9311	2 4 6	8 11 13	15 17 19
·97	9333	9354	9376	9397	9419	9441	9462	9484	9506	9528	2 4 7	9 11 13	15 17 20
·98	9550	9572	9594	9616	9638	9661	9683	9705	9727	9750	2 4 7	9 11 13	16 18 20
·99	9772	9795	9817	9840	9863	9886	9908	9931	9954	9977	2 5 7	9 11 14	16 18 20

Test Questions 1

1. What steps do you take to look after your tools?
2. What are the two main enemies of a refrigeration system?
3. What means should be adopted to prevent faults arising from these sources?
4. If a small speck of dust enters a system, how will it impair its operation?
5. An apprentice accompanies an engineer in the course of his day's work. In what ways can he be of assistance?
6. Describe how to light a Tilley leak detector. What precautions need to be taken in the process?
7. When making a 'flare' on copper pipe, mention at least two important precautions to take.
8. Why is a good customer–engineer relationship important?
9. State some of the actions which might give offence to a customer.
10. Convert: (a) 22°C into degrees Fahrenheit.
    (b) 75°F into degrees Celsius.
11. What is the difference between degree of heat and quantity of heat?
12. Define a Btu.
13. What is: (a) Latent heat?
    (b) Sensible heat?
14. Define the boiling point of a liquid.
15. How can the boiling point be varied?
16. How may a gas be liquefied?
17. How would you describe a vacuum?
18. Explain and give examples of the three methods by which heat may be transferred from one place to another.
19. How would you define the 'second law of thermodynamics'?
20. What is meant by 'ambient temperature'?
21. How many Btu must be extracted to reduce the temperature of 10 gallons of water from 90°F to 40°F? (1 gallon weighs 10 lb).
22. How many kilojoules are required to be extracted from 6 kg of water to reduce its temperature from 33°C to 3°C?

ANSWERS

Test Questions 2

1. Draw a sketch of a refrigeration system, and explain the cycle of operation.
2. What is meant by the terms 'back pressure' and 'head pressure'?
3. Explain why the 'back pressure' of a refrigerator system is of particular importance.
4. From the appropriate charts give the pressures you would expect on the head and suction gauges of a unit using R.12, operating in an ambient of 21°C (70°F).
   (a) Frozen food cabinet. (b) Coldroom operating at a temperature of 0°C (32°F). (c) Cellar cooler operating at 13°C (55°F).
5. Describe three methods of controlling the flow of liquid refrigerant in a system.
6. Sketch and describe the action of a TEV (thermostatic expansion valve).
7. Why is a TEV more adaptable than an AEV (automatic expansion valve)?
8. Enumerate the desirable qualities of the perfect refrigerant.
9. Explain the procedure necessary to fit gauges to the compressor of a refrigerator unit.
10. How would you test a compressor for pumping efficiency?
11. Describe the procedure to fit a new head plate to a compressor.
12. Describe how to renovate an inefficiently pumping head plate.
13. Describe how to change a compressor shaft seal.
14. What danger exists in a compressor: (a) With too little oil? (b) With too much oil?
15. What is the purpose of the following controls? (a) Thermostat. (b) Low-pressure control. (c) High-pressure control. (d) Water valve.
16. What is the difference in the actuating mechanism of a low-pressure control and a thermostat?
17. Why is a 'snap' action necessary on an electrical switch?

18. What is the purpose of the following? (a) Drier. (b) Sight glass. (c) Heat exchanger. (d) Non-return valve.
19. Explain with the aid of a sketch how handwheels may be of value in a system.

ANSWERS

Test Questions 3

1. State the four common measures of electricity, and show by equations how they are related.
2. Draw a sketch showing the correct connections for 'live', 'neutral' and 'earth' of an ordinary household socket.
3. Sketch and describe a solenoid.
4. Name the controls in which a solenoid is employed.
5. With the aid of sketches, give two examples of how a single-phase motor is started.
6. Sketch and describe the operation of a centrifugal switch.
7. Describe the stages and precautions necessary in stripping down a small electric motor.
8. Describe the normal operation of a domestic refrigerator.
9. Sketch, describe and explain the operations of a relay.
10. Sketch and describe an overload protector.
11. What is the purpose of a sealed unit tester?
12. A domestic refrigerator is stated to be 'running continuously'. How would you begin testing, and what are the probable causes?
13. What is meant by relative humidity?
14. What is dew point?
15. Why is the frost build-up on an evaporator undesirable?
16. Describe the methods of artificial defrosting, and sketch one system.
17. How is the automatic action achieved? Draw a sketch.
18. What are the desirable qualities of an insulating material? What is meant by the 'k factor': (a) in Imperial units, (b) in SI units?
19. What is meant by: (a) Heat load? (b) Product load? (c) Specific heat? (d) Specific heat capacity?
20. How much heat must be extracted from 5 gallons of water at 65°F, to transfer it into ice at 20°F? Take latent heat of fusion, 144 Btu.

21. How many kJ must be extracted from 20 kg of water at 20°C to transfer it into ice at −10°C? Take latent heat as 334 kJ.
22. How much heat will penetrate a coldroom the dimensions of which are 10 ft × 12 ft × 7 ft, in an ambient of 80°F, coldroom temperature 40°F? Insulation is 4 in cork, and '*k* factor' 0·3. Neglect the floor.
23. How much heat will penetrate a coldroom the dimensions of which are 3 m × 4 m × 2 m high, in an ambient of 30°C, coldroom temperature 5°C? Insulation 100 mm polystyrene with a '*k* factor' of 0·033 W/m°C. Neglect the floor.

ANSWERS

1. Page 73
2. Page 75, Fig. 32
3. Page 76
4. Pages 77–79
5. Pages 82–83, Figs. 38–39
6. Page 83, Fig. 40
7. Page 88
8. Page 96
9. Page 92, Fig. 45
10. Page 93, Fig. 46
11. Page 97, Fig. 48
12. Pages 98–99
13. Page 102
14. Page 103
15. Page 103
16. Pages 105–110, Figs. 50–53
17. Page 104, Fig. 49
18. Pages 113–114
19. Pages 116–117
20. 9 150 Btu
21. 8 770 kJ
22. 1 284 Btu/h
23. 0·33 kW

Appendix—SI Units

Whilst it is expected that all young students will be familiar with SI units, the following notes are added for the benefit of those of earlier generations. They are intended to be a guide only, and are confined to the limits of calculations contained in this book. For a proper study of SI units a special course is recommended.

SI Units for Common Quantities

		Symbol
length	metre	m
area	square metre	m^2
volume	cubic metre	m^3
mass	kilogram	kg
time	second	s
temperature	degrees Celsius	°C

SI units have one unit for each quantity, with multiples of the unit to express larger or smaller measures. The multiples most commonly used are related to the unit by factors of 1 000.

kilo (k) and mega (M) express larger measures
milli (m) and micro (μ) express smaller measures

			Prefix	Symbol
one million	1 000 000	10^6	mega	M
one thousand	1 000	10^3	kilo	k
one thousandth	0.001	10^{-3}	milli	m
one millionth	0.000 001	10^{-6}	micro	μ

Where the value is less than unity, the decimal point is preceded by zero, *e.g.* 0.123.

For weights and measures the point is placed on the line in printed, handwritten and typed material, *e.g.* 1.234, with a space used as the thousands marker. Four figures or less may be blocked together:

$$1000 \text{ mm} = 1 \text{ m}$$
$$1000 \text{ litres} = 1 \text{ m}^3$$

Five figures or more are grouped in threes divided by single space:

$$10\ 000\ \text{m}^2$$
$$1\ 000\ 000$$

When tabulating numbers, all numbers should be grouped in threes to keep them in columns:

$$1\ 000$$
$$1\ 000\ 000$$
$$10\ 000$$

The thousands marker space is used on either side of the decimal marker, *e.g.* 1.123 45.

Symbols are the same in the plural as they are in the singular, 's' being the symbol for second or seconds.

$$1\ \text{m} = \text{one metre}$$
$$10\ \text{m} = \text{ten metres}$$
$$10\ \text{ms} = \text{ten milleseconds}$$

Never put a full stop after a symbol, except at the end of a sentence. They are symbols and not abbreviations. Leave a space between figures and symbols:

$$71\ \text{m} = 71\ \text{metres}$$

To avoid confusion, write litre and tonne in full, and refer to the tonne as a 'metric tonne'.

The following is a list of units, common multiples and symbols for the quantities relating to refrigeration.

Quantity	Unit	Symbol	Multiples	Symbol
Force	newton	N	meganewton	MN
			kilonewton	kN
Energy	joule	J	megajoule	MJ
			kilojoule	kJ
Power	watt	W	megawatt	MW
			kilowatt	kW
Pressure	pascal	Pa	megapascal	MPa
			kilopascal	kPa
Current	ampere	A	milliamp	mA
Voltage	volt	V	megavolt	MV
			kilovolt	kV

Unit names written in full carry lower-case initial letters except for degree Celsius or where they begin a sentence. In general, symbols for units whose names are those of scientists are in upper case, *e.g.* A for Ampere, K for Kelvin, whilst other symbols are in lower case, *e.g.* m for metre and s for second.

The unit of force is the newton, N, which is defined as: 1 newton is the force required to give a mass of 1 kilogram an acceleration of 1 metre per second squared. In symbol form:

$$\text{force (N)} = \text{mass (kg)} \times \text{acceleration (m/s}^2)$$

The newton is a key unit in the SI system, and as the unit of force it is used to derive units for the following quantities:

Energy and work   = joule (J)
Power   = watt (W)
Pressure and stress = newton per square metre
                                = N/m^2

There are simple relationships between the newton, joule and watt:

Energy (J) = Work done   Power (W) = Rate of doing work
         = Force (N) × distance (m)
    1 J = 1 Nm   1 W = 1 J/s (1 N m/s)

One further unit of energy used in practice is the kilowatt hour

$$= 1 \text{ kW h}$$
$$1 \text{ N m} = 1 \text{ J} = 1 \text{ W s}$$

The unit of pressure is newton per square metre (N/m^2)

$$= \text{pascal (Pa)}$$
$$1 \text{ Pa} = 1 \text{ N/m}^2$$

Since the pascal is such a small unit of pressure, the multiples kilopascal and megapascal will be more commonly used in practice:

1 kilopascal = kPa = kilonewtons per square metre = kN/m^2
1 megapascal = MPa = meganewtons per square metre = MN/m^2

In addition to the pascal and its multiples, which will continue to be used in practice, are the bar, millibar (mbar) and hectobar (hbar). There is no separate symbol for the bar.

    1 bar = 100 kPa (10^5 N/m^2)
  1 mbar = 100 Pa (10^2 N/m^2)
  1 hbar = 10 000 kPa (10^7 N/m^2)

Atmospheric pressure may be expressed as 101.4 kilopascals or 1.014 bar or 1014 millibars or 760 millimetres of mercury.

Since we already use the SI units of current and voltage they will remain the same.

$$\text{Unit of current} = \text{ampere} = A$$
$$\text{milliamp} = mA$$

$$\text{Unit of voltage} = \text{volt} = V$$
$$\text{megavolt} = MV$$
$$\text{kilovolt} = kV$$

The following are a few common errors in writing SI units which are to be avoided:

Correct	Incorrect
m	m.
20 m	20m or 20ms.
mm	m.m. or mm.
m^2	sq.m or sq.m.
tonne	t.
kg	Kg. kg. or KG
m^3	cu.m.
kJ	KJ Kj or kJ.
litre	l or lit.
N	N. or n

This list is by no means complete but illustrates the variations possible, and how important it is to use the symbols in their proper form.

METRIC CONVERSION TABLE

To change:

inches	to mm	multiply by	25·4
inches	to cm	multiply by	2·54
feet	to m	multiply by	0·304 8
sq in	to sq cm	multiply by	6·452
sq ft	to sq m	multiply by	0·093
cu ins	to cu cm	multiply by	16·39
gal (US)	to litre	multiply by	3·785
gal (Imp)	to litre	multiply by	4·55
lb	to kg	multiply by	0·454
oz av	to g	multiply by	28·35
Btu	to kcal	multiply by	0·252
Tons refrig	to kcal	multiply by	3 024·0
pint (US)	to litre	multiply by	0·473
pint (Br)	to litre	multiply by	0·568
psi	to kg sq cm	multiply by	0·07
Btu h sq ft °F	to kcal h sq m °C	multiply by	4·88
mm	to inches	multiply by	0·039 4
cm	to inches	multiply by	0·394
m	to inches	multiply by	39·37
m	to feet	multiply by	3·281
sq cm	to sq ins	multiply by	0·155
sq m	to sq ft	multiply by	10·76
cu cm	to cu ins	multiply by	0·061
cu m	to cu ft	multiply by	35·32
litre	to gal (Imp)	multiply by	0·22
litre	to gal (US)	multiply by	0·264
kg	to lb	multiply by	2·205
g	to oz av	multiply by	0·352
kcal	to Btu	multiply by	3·97
litre	to pint (US)	multiply by	2·11
litre	to pint (Br)	multiply by	1·76
kg cu cm	to psi	multiply by	14·22
kcal h sq m °C	to Btu h sq ft °F	multiply by	0·205
kcal	to kJ	multiply by	4·18
Btu	to kJ	multiply by	1·053
kJ	to kcal	multiply by	0·24
kJ	to Btu	multiply by	0·95

Index

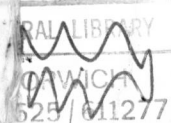